MATHEMATICS

For the HiSET® Test

New Readers Press
ProLiteracy's publishing division

Photo courtesy of:
p. 188: © Flashon Studio

Mathematics for the HiSET® Test
ISBN 978-1-56420-885-9

New Readers Press
ProLiteracy's Publishing Division
101 Wyoming Street, Syracuse, New York 13204
www.newreaderspress.com

Printed in the United States of America
10

Proceeds from the sale of New Readers Press materials support professional development, training, and technical assistance programs of ProLiteracy that benefit local literacy programs in the U.S. and around the globe.

Developer: QuaraCORE
Editorial Director: Terrie Lipke
Cover Design: Carolyn Wallace
Technology Specialist: Maryellen Casey

Contents

Using This Book

Welcome to *Mathematics for the HiSET® Test*, an important resource in helping you build a solid foundation of mathematic skills as you prepare for the HiSET® high school equivalency test.

- First, read *What to Expect* on page 5, which will give you a brief overview of the HiSET® test itself.
- Next, take the Pretest, which begins on page 6. After taking the Pretest and checking your answers, use the chart on page 13 to find the lessons that will help you study the skills you need to improve.
- Then, start using the book, which is organized into seven units, each containing brief lessons that focus on specific skills. Important math terms included in the unit are listed on the first page of the unit and appear in **boldface** when they are first used in each lesson. Use the Glossary at the end of each unit to find key term definitions.
- Along the way, you can use the Key Math Formulas on page 184 as a reference to help you solve problems.
- Each lesson is followed by a Lesson Practice with questions to test your knowledge of the lesson content. Answers can be found in the Answer Key at the end of the unit. Each Lesson Practice also includes *Key Point!* and *Test Strategy* tips to help you prepare for the HiSET® test.
- Each unit concludes with a Unit Test that covers all the content in the unit's lessons. The Unit Test Answer Key appears at the end of each unit.
- Every unit concludes with *Study More!*, which lists additional skills and topics you can study to prepare for the HiSET® test.
- After completing all the units, you can test what you know by taking the HiSET® Practice Test, beginning on page 174. This test will help you check your understanding of all the skills in the book.

What to Expect

This book is intended to help you prepare to take the HiSET® (High School Equivalency Test) Exam in Mathematics. This is one of the five HiSET® exams; the others are in Reading, Writing, Science, and Social Studies. The HiSET® exams are available in English and in Spanish and can be taken in written format (on paper) or on a computer. For more information about the HiSET® exams, visit hiset.ets.org/test_takers.

Preparing for the Test

- Using *Mathematics for the HiSET® Test* is a great first step in preparing to take the test. This book provides an overview of the content of the Mathematics test and gives you many opportunities to take practice tests to evaluate how well you know the content and skills that will be included in the HiSET® test.
- Be sure to allow enough time to prepare for the test. Choose a test date that will not force you to rush through your study period. You need time to use this book, and possibly more time for additional review and practice tests.
- When using this book, the Pretest helps you identify your strengths and weaknesses in math skills so you can pinpoint the areas on which to focus your study for the HiSET®.
- Be sure to use the Key Formulas page at the end of this book so you learn them prior to taking the HiSET® test. Some, but not all, of these formulas will be provided for you when you take the test.

Taking the Mathematics Test

The HiSET® Mathematics exam includes 55 multiple-choice questions. You will have 90 minutes to complete the test. A four-function calculator will be provided for you at the test center.

The Mathematics exam will assess your understanding of key math skills taught in high school, including:

- numbers and operations
- measurement
- geometry
- data analysis
- probability
- statistics
- algebraic concepts

These and many other math skills are reviewed in this book. Each lesson covers a specific skill, and each is followed by practice questions to help you learn the content. If you then complete the practice HiSET® test at the end of this book, you can judge your results and decide if you are ready to take the HiSET® Mathematics test.

Answer the following questions to gauge your readiness to take the HiSET® test. Answers to questions appear on page 12.

Questions 1 and 2 refer to the coordinate grid below.

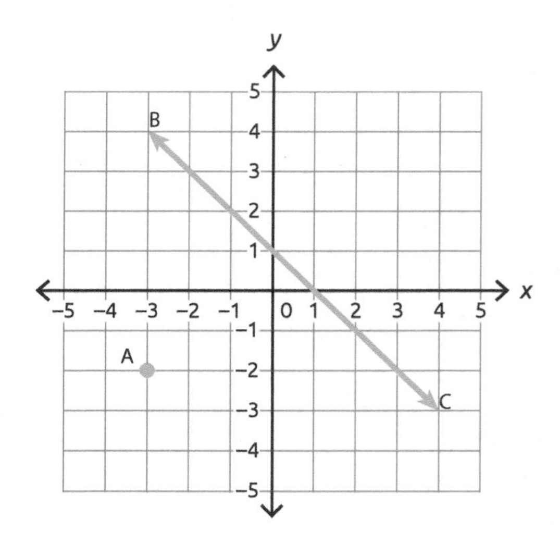

1. What is the coordinate location of point A?
   A (−3, 2)
   B (3, 2)
   C (−3, −2)
   D (3, −2)
   E (−2, −3)

2. What is the y-intercept of $\overleftrightarrow{BC}$?
   A (0, 1)
   B (1, 0)
   C (0, −1)
   D (−1, 0)
   E (1, 1)

3. Nora is playing chess and slides her queen diagonally three places. Which transformation does Nora perform with her queen?
   A translation
   B reflection
   C rotation
   D dilation
   E reproduction

Questions 4 and 5 refer to the information below.

A sporting goods store took a survey to determine the favorite professional sports leagues of its customers. Questions 4 and 5 refer to the results of the survey, shown in the graph.

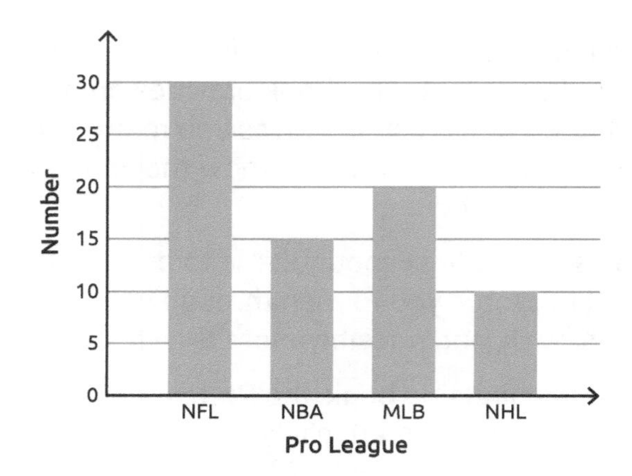

4. How many people like the NFL or MLB the best?
   A 20
   B 30
   C 45
   D 50
   E 75

5. About what percentage of customers like the NBA best?
   A 50%
   B 40%
   C 20%
   D 15%
   E 10%

Questions 6 and 7 refer to the figure shown below.

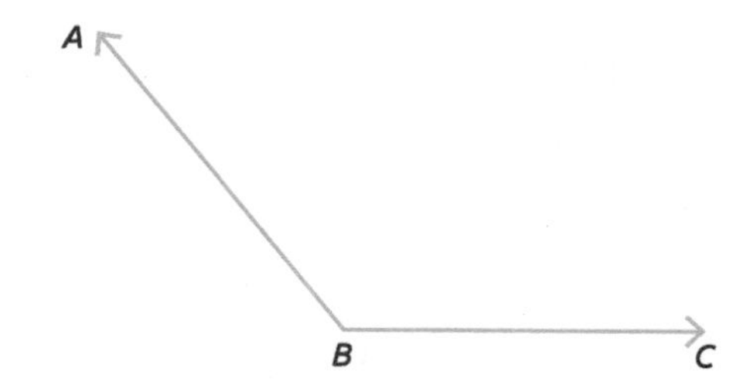

6. In the figure, what is ABC called?
   - **A** angle
   - **B** point
   - **C** ray
   - **D** line
   - **E** line segment

7. In the figure, what is $\overrightarrow{BC}$ called?
   - **A** point
   - **B** angle
   - **C** line
   - **D** line segment
   - **E** ray

8. Samantha jumps $3\frac{1}{5}$ feet in the standing long jump. Her goal is to jump $4\frac{1}{4}$ feet. How much farther does she need to jump to meet her goal?
   - **A** $1\frac{1}{20}$ feet
   - **B** $1\frac{1}{5}$ feet
   - **C** $1\frac{1}{4}$ feet
   - **D** $1\frac{1}{2}$ feet
   - **E** $1\frac{3}{4}$ feet

9. G is the midpoint of $\overline{FH}$ in the following diagram. If $\overline{FH} = 28$, what is the value of x?

   F ——— $2x + 2$ ——— G ——— $3x - 4$ ——— H

   - **A** 2
   - **B** 6
   - **C** 8
   - **D** 16
   - **E** 32

10. Gary earns 3% commission for each home he sells. What does he earn in a month in which he sells $325,000 worth of homes?
    - **A** $325
    - **B** $975
    - **C** $9,750
    - **D** $97,500
    - **E** $975,000

11. On the first night of a concert, 5,500 people attended. The next night, the attendance tripled. Which expression represents the total attendance for the first two nights of the concert?
    - **A** $5{,}500 \times (3 \times 5{,}500)$
    - **B** $5{,}500 - (3 \times 5{,}500)$
    - **C** $5{,}500 + (3 \times 5{,}500)$
    - **D** $(3 \times 5{,}500) - 5{,}500$
    - **E** $(3 \times 5{,}500) \div 5{,}500$

12. Fredia has four pieces of yarn. The pieces of yarn have lengths of 9 inches, 14 inches, 26 inches, and 11 inches. What is the total length of all four pieces of yarn?
    - **A** 2 feet
    - **B** 4 feet
    - **C** 5 feet
    - **D** 6 feet
    - **E** 8 feet

13. Admission for a play is \$9 for adults and \$5 for students. The first night, the box office sold twice as many adult tickets as student tickets. The second night, the number of each type of ticket sold was three times the number sold the first night. Which expression can be used to determine the total amount of money received for sales of tickets on both nights? Let *a* be the number of adult tickets sold and *s* be the number of student tickets sold.

   A $3(9a + 5s) + 3(9a + 5s)$
   B $3 \cdot 9a + 5s$
   C $3(9a + 5s)$
   D $3(9a + 5s) + 9a + 5s$
   E $3(9a + 5s) - 9a + 5s$

14. Evaluate the expression $5a + b$ if $a = -3$, and $b = 4$.

   A −19
   B −11
   C −4
   D 11
   E 19

15. A high school has 800 students. The ratio of boys to girls is 2 to 3. How many students are boys?

   A 267
   B 320
   C 400
   D 480
   E 533

16. Which of the following demonstrates the Commutative Property for addition?

   A $(4 + 11) + 2 = 4 + (11 + 2)$
   B $(4 \times 11)2 = 4(11 \times 2)$
   C $4 + 11 + 2 = 11 + 4 + 2$
   D $4 \times 11 \times 2 = 2 \times 4 \times 11$
   E $4(11 + x) = 44 + 4x$

17. Which number line shows the solution to $-x - 2 + 5x - 10 \geq 0$?

   A

   B

   −5 −4 −3 −2 −1 0 1 2 3 4 5

   C

   −5 −4 −3 −2 −1 0 1 2 3 4 5

   D

   −5 −4 −3 −2 −1 0 1 2 3 4 5

   E
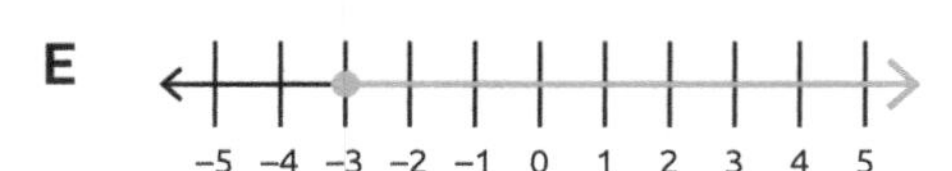

18. The area of a circle is found using the formula $A = \pi r^2$, where *r* is the radius of the circle. What is the area of the circle whose radius is 10 inches? Use $\pi = 3.14$.

   A 3.14 in^2
   B 31.4 in^2
   C 314 in^2
   D 3,140 in^2
   E 31,400 in^2

19. Solve by factoring $x^2 + 6x - 7 = 0$.

   A $\{-7, 1\}$
   B $\{-1, 7\}$
   C $\{-7, -1\}$
   D $\{1, 7\}$
   E $\{1, 0, -7\}$

20. If $f(x) = -2x + 1$, what is $f(-5)$?

   A −11
   B −9
   C −5
   D 9
   E 11

21. The graph of a function is shown here. What type of function is it?

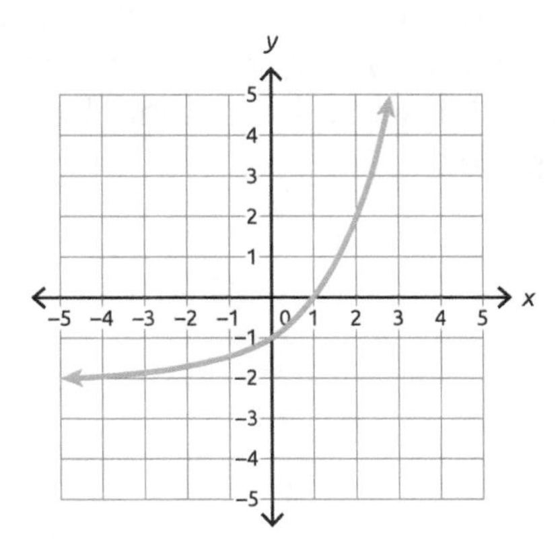

A linear
B quadratic
C exponential
D absolute value
E piecewise

22. Kiah drives to the beach for vacation, and it takes her 6 hours to get there. Which of the following distances is most likely the distance she drove?

A 50 miles
B 100 miles
C 300 miles
D 900 miles
E 1,000 miles

23. What is the difference in the expression shown here?

$(5a^2 - 2a + 1) - (3a^2 - 5a + 4)$

A $2a^2 + 3a + 3$
B $2a^2 + 3a - 3$
C $2a^2 - 3a + 3$
D $8a^2 + 7a + 5$
E $8a^2 - 8a - 3$

24. Which of the function types does the following table represent? Explain.

x	f(x)
1	5
2	8
3	11
4	14
5	17

____

____

____

25. In 2014, about 12% of the U.S. population lived in California. Which shows the ratio of Californians to the total U.S. population in 2014?

A 6:25
B 3:25
C 4:5
D 1:12
E 15:16

26. If $m\angle T = 99°$, what type of angle is $\angle T$?

A acute
B right
C obtuse
D straight
E supplementary

27. Which of the following numbers is equivalent to $\sqrt{48}$?
   - A 7
   - B $4\sqrt{6}$
   - C $3\sqrt{16}$
   - D $16\sqrt{3}$
   - E $4\sqrt{3}$

28. Which type of function forms a parabola when graphed on a coordinate plane?
   - A linear
   - B quadratic
   - C exponential
   - D absolute value
   - E piecewise

29. The height of a mature apple tree is 12 units tall. Which of the following is the most reasonable unit for the height of the tree?
   - A inches
   - B feet
   - C miles
   - D kilometers
   - E centimeters

30. For a game, Ethan spins a spinner that has 6 equal-sized sections, each with a different color. The spinner uses the following colors: red, blue, green, yellow, purple, and orange. What is the probability that the spinner lands on the yellow space?
   - A $\frac{1}{2}$
   - B $\frac{1}{3}$
   - C $\frac{1}{4}$
   - D $\frac{1}{6}$
   - E $\frac{1}{8}$

31. What is the solution of the system of equations shown below?

$$\begin{cases} x + y = 10 \\ x - 2y = 1 \end{cases}$$

   - A (6, 4)
   - B (3, 7)
   - C (7, 3)
   - D There is no solution.
   - E There are infinitely many solutions.

32. A company's profit for the last 4 months is shown in the table. What was the net profit during these months?

Month	Profit ($)
April	11,575
May	−4,325
June	−3,120
July	6,090

   - A −$11,960
   - B $10,220
   - C $16,460
   - D $17,665
   - E $25,110

33. A fruit punch recipe requires $3\frac{1}{4}$ ounces of sugar for each gallon of punch made. Which of the following would give the same concentration of sugar for the punch?
   - **A** 10 ounces of sugar for 3 gallons of punch
   - **B** 13 ounces of sugar for 4 gallons of punch
   - **C** $6\frac{3}{4}$ ounces of sugar for 2 gallons of punch
   - **D** $1\frac{1}{2}$ ounces of sugar for $\frac{1}{2}$ gallon of punch
   - **E** $\frac{1}{2}$ ounces of sugar for $\frac{1}{4}$ gallon of punch

34. Kyle is the tallest guy on his basketball team at 77 inches tall. How tall is Kyle?
   - **A** 6 feet
   - **B** 6 feet 2 inches
   - **C** 6 feet 5 inches
   - **D** 6 feet 8 inches
   - **E** 7 feet 7 inches

35. Phillip flips a quarter and a nickel at the same time. How many possible outcomes are there for flipping two coins?
   - **A** 1
   - **B** 2
   - **C** 4
   - **D** 5
   - **E** 6

36. What is the measure of the angle that is supplementary to $\angle ABC$ shown below?

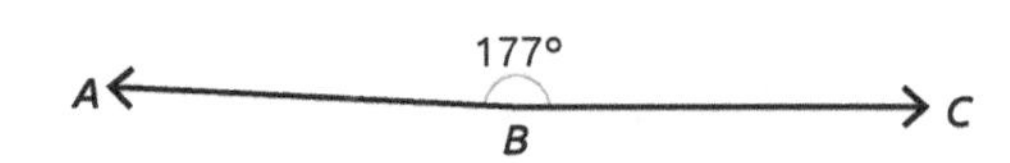

   - **A** 87°
   - **B** 23°
   - **C** 13°
   - **D** 3°
   - **E** 1°

Use the following information to answer questions 37 and 38.

Jennifer's race times for the season so far are shown in the following table.

Race	Time (sec)
1	55
2	58
3	49
4	65
5	51

37. What is the mean time, in seconds, for all of her races?
   - **A** 14
   - **B** 52.8
   - **C** 55
   - **D** 55.6
   - **E** 65

38. What is the median time, in seconds, for all of her races?
   - **A** 14
   - **B** 52.8
   - **C** 55
   - **D** 55.6
   - **E** 65

39. A retail store marks up its merchandise 40% above the wholesale price. If the store sells shoes for $75, which equation can be used to determine x, the wholesale price of the shoes in dollars?
   - **A** $x + 0.40 = 75$
   - **B** $x - 0.40 = 75$
   - **C** $1.40x = 75$
   - **D** $0.60x = 75$
   - **E** $0.40x = 75$

Answer Key

1. **C.**
2. **A.** The y-intercept is where the line crosses the y-axis. The line crosses the y-axis at (0, 1).
3. **A.**
4. **D.** Add the number who like the NFL, 30, and MLB, 20, together.
5. **C.** Divide the number who like the NBA best, 15, by the total number surveyed, 75.
6. **A.**
7. **E.**
8. **A.** Subtract $4\frac{1}{4} - 3\frac{1}{5}$ to find the additional distance needed.
9. **B.**
10. **C.** Multiply $325,000 by 0.03.
11. **C.**
12. **C.**
13. **D.** The money brought in the first night was 9 times the number of adult tickets plus 5 times the number of student tickets, or $9a + 5s$. The second night was three times this amount, or $3(9a + 5s)$. Add these two amounts: $3(9a + 5s) + 9a + 5s$.
14. **B.** Substitute −3 for a and 4 for b and simplify. $5(-3) + 4 = -15 + 4 = -11$.
15. **B.** Two out of five students are boys. Two-fifths of 800 is 320.
16. **C.**
17. **C.** Simplify the expression to $x \geq 3$. The graph of this inequality is a closed circle on 3 with a ray going to the right.
18. **C.**
19. **A.** Factor and set both factors equal to zero. Then, solve for x in both equations.
20. **E.** Substitute −5 for x in the expression $-2x + 1$ and simplify.
21. **C.**
22. **C.**
23. **B.**
24. **It is a linear function. Both columns keep adding the same amount: as the *y*-column increases by 1, the *x*-column increases by 3. This means the rate of change is constant, which is true of linear functions.**
25. **B.** Percent is out of 100. Simplify the ratio.
26. **C.** An angle with a measure greater than 90° and less than 180° is an obtuse angle.
27. **E.**
28. **B.**
29. **B.**
30. **D.**
31. **C.**
32. **B.**
33. **B.**
34. **C.**
35. **C.**
36. **D.** Supplementary angles add up to 180°. $180° - 177° = 3°$.
37. **D.**
38. **C.**
39. **C.**

Check your answers. Review the questions you did not answer correctly. You can use the chart below to locate lessons in this book that will help you learn more about math content and skills. Which lessons do you need to study? Work through the book, paying close attention to the lessons in which you missed the most questions. At the end of the book, you will have a chance to take another test to see how much your score improves.

Question	Where to Look for Help		
	Unit	**Lesson**	**Pages**
1, 2	3	4	66–67
3	5	3	118
4, 5	7	2	157
6, 7, 9, 26	5	1	109–110
8	1	2	21
10, 15, 25, 33, 39	1	4	26–27
11, 13, 16, 32	2	1	37–38
12, 29	6	1	132–133
14	3	1	53
17	3	3	62–63
18, 36	5	2	112–113
19, 23	3	2	57–58
20	4	1	95–96
21, 24, 28	4	2	99
22	1	1	15–16
27	3	7	80
30, 35	7	4	164–165
31	3	5	71
34	6	3	141
37, 38	7	1	151–152

UNIT 1

Numbers and Operations; Ratios

Operations with numbers occur in many places and in various forms every day. Consider the following scenario:

Sharon earns $15.25 per hour at her maintenance job. Her time card for last week is shown. What are her total earnings for the week? And if she wants to put 15% of her earnings into a savings account, how much money will she deposit from last week's earnings?

Answer: $503.25; about $75.49

Time Card	
Day:	**Hours:**
Monday	6.25
Tuesday	7.5
Wednesday	6.75
Thursday	7
Friday	5.5

KEY WORDS

- commission
- integer
- irrational numbers
- markdown
- markup
- natural numbers
- percent
- place value
- proportion
- radical number
- ratio
- rational number
- real number
- simple interest
- unit rate
- whole numbers

Real Number System

The **real number** system consists of any number that can be represented on a number line. There are several subsets of real numbers, as shown in this Venn diagram.

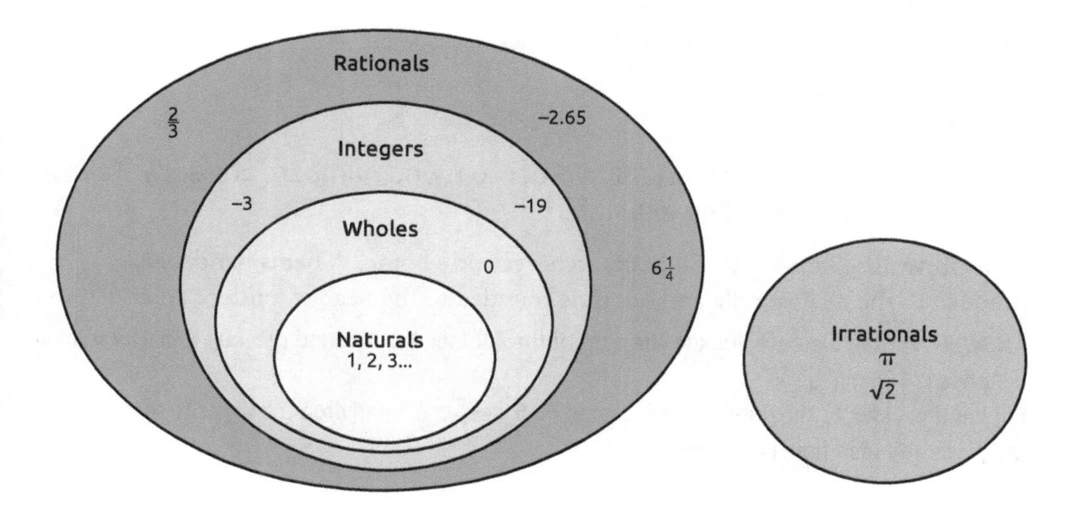

In general, real numbers can be classified as rational numbers or irrational numbers. **Rational numbers** are numbers that can be written as the fraction of two integers. Rational numbers include **natural numbers** (1, 2, 3, ...), **whole numbers** (0, 1, 2, ...), **integers** (..., -2, -1, 0, 1, 2, ...), all fractions, and all repeating or terminating decimals. **Irrational numbers** include any non-terminating, non-repeating decimals as well as square roots of numbers that are not perfect squares.

Place Value

When reading and writing numbers, each digit has a value called a **place value**. Review the following chart, which demonstrates the place value of several digits on each side of the decimal point.

hundred millions	ten millions	millions	hundred thousands	ten thousands	thousands	hundreds	tens	ones	decimal	tenths	hundredths	thousandths	ten thousandths
									•				

Understanding the Real Number System

SKILLS TIP

To write a number in expanded form, multiply each digit by its place value. This form demonstrates the ' of each digit. For example,

$452.19 =$

$4 \times 100 + 5 \times 10 + 2 \times 1 +$
$1 \times (\frac{1}{10}) + 9 \times (\frac{1}{100})$

Look at each number below and identify the value of the 5 in each number.

4,572 23.537 375.2983 4,372,980.615

ANSWER: In 4,572, the 5 is in the hundreds place, making the value five hundred. In 23.537, the 5 is to the right of the decimal point and is in the tenths place. So, the value of the 5 is five tenths. In 375.2983, the 5 is in the ones place making the value simply five. In 4,372,980.615, the 5 is in the thousandths place. So, the value of the 5 is five thousandths.

Rounding

Identifying place value is important in order to round numbers correctly. Take a look at the following rounding example.

Sally drove 124.8 miles in 2.34 hours to her friend's home. What is her distance rounded to the nearest mile and her time rounded to the nearest tenth of an hour?

ANSWER: For the distance, look to the right of the 4. Since $8 > 5$, round the 4 up to 5. Her rounded distance is 125 miles.
For her time, look to the right of the 3. Since $3 < 5$, keep the 3 and drop the digits to the right of the 3. Her rounded time is 2.3 hours.

Analyzing Rounded Numbers

Rounded numbers are close to their actual values. By analyzing rounded numbers, valuable information can be determined about the actual values. The rule used for rounding numbers can be useful when describing their actual values.

Amaress has raised approximately $800 to donate to charity. What range describes the actual amount of money that Amaress likely raised?

ANSWER: The actual amount that Amaress raised has been rounded to $800. Since numbers can be rounded up or down, the actual amount could be greater or less than $800. Since 800 has been rounded to the nearest hundred, select a range of values that will round to 800 when rounded to the nearest hundred: Amaress likely raised between $750 and $849.

Comparing Numbers

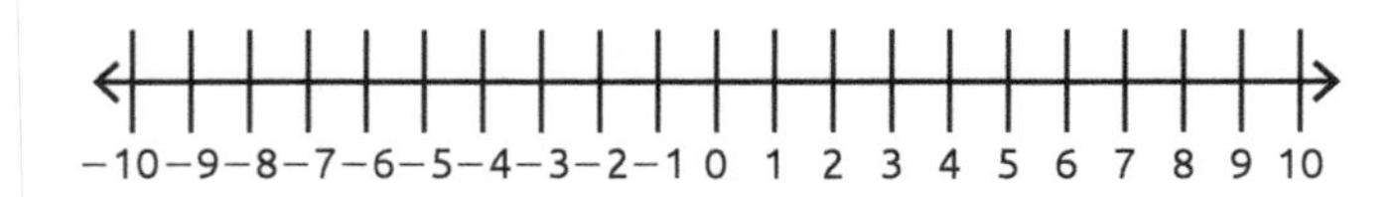

A number line can be used to compare numbers. A number is greater than another number if it lies to the right of the number on the number line. It is less than the number if it lies to the left of the number.

Use the number line above to compare the following numbers. Insert < or >.

5 _____ 3 −2 _____ −4 −7 _____ 0

ANSWER: $5 > 3$ $-2 > -4$ $-7 < 0$

MATH FACT

Perfect squares simplify to whole numbers when you compute their square roots. It is helpful to memorize the first several perfect squares.

$2^2 = 4$	$3^2 = 9$	$4^2 = 16$	$5^2 = 25$	$6^2 = 36$	$7^2 = 49$
$8^2 = 64$	$9^2 = 81$	$10^2 = 100$	$11^2 = 121$	$12^2 = 144$	$13^2 = 169$

Radicals

A **radical number** has a radical sign, $\sqrt{\ }$, over the number. It is also called a square root. To determine the square root of a number, find the whole number that when multiplied by itself results in the number under the radical. For example, $\sqrt{4} = 2$ because $2 \times 2 = 4$.

If the number under the square root is a perfect square, then it is a rational number. If it is not a perfect square, then it is an irrational number.

Determine if the radicals shown below are rational or irrational. Explain.

$\sqrt{10}$ $\sqrt{49}$

ANSWER: $\sqrt{10}$ is irrational because 10 is not a perfect square. $\sqrt{49}$ is a rational number because 49 is a perfect square. $\sqrt{49} = 7$.

Complete the activities below to check your understanding of the lesson content.

1. Which of the following numbers is irrational?

   A -18

   B $\sqrt{36}$

   C $\frac{2}{3}$

   D 2.875

   E $\sqrt{27}$

2. Which number is 31,092.4758 rounded to the nearest ten?

   A 3,109

   B 31,090

   C 31,092

   D 31,092.4

   E 31,092.5

Lesson Practice

TEST STRATEGY

When you are asked which statement is false, look for a counterexample. A counterexample is an example that proves that a statement is *not* true. For example: The product of any two integers is negative if one of the integers is negative. This is a counterexample: $0 \times -2 = 0$

3. Arturo is climbing to the top of a mountain that is about 6,000 ft tall. Which shows the most likely range for the actual height of the mountain?
   A. between 5,000 ft and 7,000 ft
   B. between 5,500 ft and 6,000 ft
   C. between 5,500 ft and 6,500 ft
   D. between 6,000 ft and 6,500 ft
   E. between 6,000 ft and 8,000 ft

4. Which of the following statements is true?
   A The number 12,389.75 rounded to the tens place is 12,389.8.
   B The number 1,371.673 rounded to the tenths place is 1,370.
   C The number 895.62 rounded to the ones place is 896.
   D The number 7,256.876 rounded to the hundreds place is 7,200.
   E The number 567.239 rounded to the tens place is 567.

5. Which of the following statements is false?
   A All real numbers are either rational or irrational.
   B All integers are whole numbers.
   C A non-terminating, repeating decimal number is rational.
   D An integer is rational.
   E The set of whole numbers has one more member than the set of natural numbers.

6. The Hurricanes won their golf match. This table shows the score of each player in comparison to par for the course. Which answer choice places the players, in order of shots, from least to greatest?

Player	Total Shots
Ed	−11
Dequan	−6
Chris	−8
Payton	−2

A Dequan, Ed, Chris, Payton
B Payton, Dequan, Chris, Ed
C Ed, Chris, Dequan, Payton
D Ed, Dequan, Chris, Payton
E Payton, Chris, Dequan, Ed

7. According to the points on this number line, which inequality is NOT true?

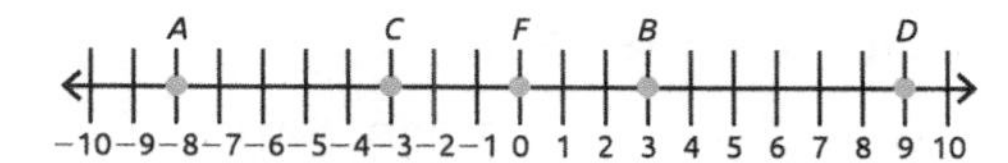

A $A > D$
B $C > A$
C $F > C$
D $B > F$
E $C < B$

8. Which of the following radicals is an irrational number?

A $\sqrt{81}$
B $\sqrt{64}$
C $\sqrt{40}$
D $\sqrt{9}$
E $\sqrt{1}$

See page 33 for answers and help.

Performing Operations on Real Numbers

SKILLS TIP

A fraction must be written in lowest terms. If the numerator and denominator can be divided by the same number, then the fraction can be reduced. For example:

$$\frac{10}{12}=\frac{10\div 2}{12\div 2}=\frac{5}{6}$$

So, $\frac{10}{12}$ can be reduced to $\frac{5}{6}$.

Addition

When adding whole numbers or decimal numbers, the numbers must be lined up by place value before adding. For decimals, line up the numbers at the decimal point and then add.

Add: $3{,}720+405$ $\qquad$ $0.285+6.03$

ANSWER:

$$\begin{array}{r} 3{,}720 \\ +\ \ 405 \\ \hline 4{,}125 \end{array} \qquad \begin{array}{r} 0.285 \\ +\,6.030 \\ \hline 6.315 \end{array}$$

To add fractions, you must have the same, or common, denominator for each fraction. Once there is a common denominator, add the numerators and keep the same denominator.

Add: $\frac{2}{3}+\frac{1}{4}$

ANSWER: $\frac{2}{3}+\frac{1}{4}=\frac{2\times 4}{3\times 4}+\frac{1\times 3}{4\times 3}=\frac{8}{12}+\frac{3}{12}=\frac{11}{12}$

Subtraction

Subtracting whole numbers and decimals is similar to adding: first line up the numbers by place value, and then subtract. Look at the following problem.

Lisa orders a sandwich that costs $7.65. She pays with a $10 bill. How much change does she receive in return?

ANSWER:

$$\begin{array}{r} 10.00 \\ -\,7.65 \\ \hline 2.35 \end{array}$$

Subtracting fractions is also similar to adding fractions. First, find a common denominator, and then subtract the numerators while keeping the same denominator.

Lem buys $\frac{3}{4}$ of a pound of candy. Jesse buys $\frac{1}{3}$ of a pound of candy. How much more candy does Lem have?

ANSWER: $\frac{3}{4}-\frac{1}{3}=\frac{3\times 3}{4\times 3}-\frac{1\times 4}{3\times 4}=\frac{9}{12}-\frac{4}{12}=\frac{5}{12}$

Multiplication

To multiply numbers with decimals, find the product of the numbers while disregarding the location of the decimal points. After multiplying the numbers, count the number of digits that are after the decimal point in each of the numbers. Next, count the same number of places over from the right to the left in the product, and then insert a decimal point. Look at the following problem.

Petra earns $12.85 per hour. How much will she earn if she works 6.5 hours?

ANSWER:

$$\begin{array}{r} 12.85 \\ \times \quad 6.5 \\ \hline 6425 \\ +77100 \\ \hline 83.525 \end{array}$$

So, Petra earns $83.53.

To multiply fractions, simply multiply straight across (numerator by numerator and denominator by denominator), and then reduce the product, if possible.

Multiply: $\frac{6}{7} \times \frac{4}{5}$

ANSWER: $\frac{6}{7} \times \frac{4}{5} = \frac{24}{35}$

Division

To divide numbers with decimals, you must move the decimal point to the right of the ones place in the divisor. In order to do this, the decimal point must also be moved the same number of places to the right in the dividend as the number of places the decimal point was moved in the divisor. Then, proceed with the long division.

Divide: 75.9 ÷ 5.75

ANSWER: $5.75\overline{)75.9}$

$$\begin{array}{r} 13.2 \\ 575\overline{)7590.0} \\ -575 \\ \hline 1840 \\ -1725 \\ \hline 1150 \\ -1150 \\ \hline 0 \end{array}$$

To divide fractions, leave the first one, then flip the second fraction, and then change the operation to multiplication. In other words, multiply by the reciprocal of the second fraction.

Divide: $\frac{4}{7} \div \frac{3}{5}$

ANSWER: $\frac{4}{7} \div \frac{3}{5} = \frac{4}{7} \times \frac{5}{3} = \frac{20}{21}$

Performing Operations on Real Numbers

Positive and Negative Numbers

When adding positive and negative numbers, two rules apply:

1. If the signs are the same, both positive or negative, add the numbers and keep the sign.
2. If the signs are different, one positive and one negative, subtract the numbers and keep the sign of the larger number.

The current temperature in an Alaskan city is −8°F. During the afternoon, the temperature rises 12°F. What is the temperature in the afternoon?

ANSWER: $-8 + 12 = 4$°F

When multiplying or dividing positive and negative numbers, two rules apply:

1. If the signs are the same, both positive or negative, the product/quotient is positive.
2. If the signs are different, one positive and one negative, the product/quotient is negative.

Each minute, the submarine dives another 15 feet below sea level. After 7 minutes, what is the depth of the submarine in relation to sea level?

ANSWER: $-15 \times 7 = -105$ ft

UNIT 1 / LESSON 2

Lesson Practice

Complete the activities below to check your understanding of the lesson content.

1. Kelvin has 24 friends over to his house to watch the championship game. Three-fourths of his friends want the home team to win. How many of his friends want the home team to win?

   A 6
   B 16
   C 18
   D 20
   E 24

2. Malia had $327.58 in her checking account. She wrote checks for $78.55 and $129.14. She also made a deposit of $178.32. What is the balance in her checking account?

   A −$58.43
   B $298.21
   C $327.58
   D $356.95
   E $713.59

3. Jesse is building benches to sell at the fair. Each bench costs him $30 to build. He is making 25 benches to sell and will sell them for $75 each. How many benches does Jesse need to sell in order to cover the cost of making all of them?
   - **A** 3
   - **B** 5
   - **C** 10
   - **D** 15
   - **E** 25

4. During the week, Ana feeds $4\frac{1}{2}$ packages of crackers to the children in her daycare class each day. On Saturdays, she feeds the children half of what she feeds them each weekday. How many of packages of crackers does she feed the children each Saturday?
   - **A** $6\frac{3}{4}$
   - **B** 4
   - **C** $2\frac{1}{2}$
   - **D** $2\frac{1}{4}$
   - **E** 2

5. Paul pays his brother $2.50 each week toward the amount that he owes him. If he owes his brother $33.75, how many weeks will it take him to pay off the amount he owes his brother?
   - **A** 13
   - **B** 14
   - **C** 15
   - **D** 36
   - **E** 37

Questions 6 and 7 refer to the information in the table.

This table shows the average monthly income of several types of employees within a school.

Employee	Average Monthly Income
Teacher's Aide	$2,400
Maintenance	$2,600
Teacher	$3,500
Principal	$4,200

6. Noelle is a teacher in this school. She has $2,650 in monthly living expenses. She also has a car loan for $9,000. In how many months could Noelle pay off her car loan?
   - **A** 3
   - **B** 4
   - **C** 10
   - **D** 11
   - **E** 12

7. How much more money will a teacher make at this school, on average, than a teacher's aide in a year?
   - **A** $13,200
   - **B** $10,800
   - **C** $8,400
   - **D** $2,400
   - **E** $1,100

See page 33 for answers and help.

Describing a Ratio

Ratios

A **ratio** is simply a comparison of two numbers. Ratios are commonly written in word form, 2 to 3, with a colon, 2:3, or as a fraction, $\frac{2}{3}$. Look at the following example of how ratios might be used.

A soccer team can have 11 players on the field at one time. One formation has 2 forwards, 5 midfielders, 3 defenders, and 1 goalkeeper. What is the ratio of midfielders to the total number of players that a team can have on the field at one time?

ANSWER: $\frac{\text{Midfielders}}{\text{Total Players}} =$ or 5:11, or 5 to 11

Unit Rate

A ratio can also be thought of as a rate. A **unit rate** compares an amount of one thing to a single amount of another thing, to a quantity of one. For example, if a tub is filling at a rate of 10 gallons per 2 minutes, the unit rate is 5 gallons per 1 minute. Unit rates make it easier to compare different rates. Look at the next example.

A store sells soup in three different-sized cans. The first is an 8-ounce can that sells for $0.96, the second is a 20-ounce can that sells for $2.15, and the third is a 32-ounce can that sells for $3.76. Find the unit price for each can. Which can is the cheapest to buy in terms of cost per ounce?

ANSWER: $\frac{0.96}{8} = \frac{0.12}{1}$ or, 12 cents per oz; $\frac{2.15}{20} = \frac{.1075}{1}$ or, 10.75 cents per oz; $\frac{3.76}{32} = \frac{.1175}{1}$ or, 11.75 cents per oz.

The 20-ounce can is the cheapest to buy since it has the lowest unit price.

Unit rates can be used to find other costs as well.

A bunch of grapes weighs 4 pounds and costs $7.56. Using this rate, how much would 7 pounds of grapes cost?

ANSWER: Unit Rate: $\frac{7.56}{4} = \frac{1.89}{1}$ or $1.89 per lb;

Cost of 7 pounds: $\$1.89 \times 7 = \$13.23$

Complete the activities below to check your understanding of the lesson content.

Consider the ad below.

1. Which expression represents the cost of 375 pounds of meat?

   A $\frac{375}{40} \times 15$

   B $\frac{15}{375} \times 40$

   C $\frac{15}{40} \times 375$

   D $\frac{375}{15}$

   E $\frac{375}{15} \times 40$

2. A grocery store is having a sale on crackers. A customer can buy 3 boxes of crackers for $5. Using this rate, what would 5 boxes of crackers cost?

   A $1.33

   B $1.50

   C $7.50

   D $8.33

   E $15

3. A spice company offers different prices depending on how many ounces of each spice a customer buys. Which of the following is the most affordable spice to buy per ounce?

   A 22 ounces for $2.75

   B 11 ounces for $1.30

   C 30 ounces for $2.93

   D 19 ounces for $1.98

   E 25 ounces for $2.25

See page 33 for answers and help.

Using Ratios and Proportions to Solve Problems

Proportions

A **proportion** is simply two ratios that are equal to each other. A proportion can be used in many real-world situations. An unknown value in a proportion can be solved for by using cross-multiplication. Look at the following situation and use a proportion to solve it.

If a factory can manufacture 84 bicycles per day when it is at 70 percent capacity, then how many bicycles can it produce at 100 percent capacity?

ANSWER: $\frac{84}{70} = \frac{b}{100}$

$70b = 84 \times 100$

$70b = 8{,}400$

$b = 120$ *bicycles*

Simple Interest

Simple interest is interest that is paid to a lender and is computed each year based solely on the original loan amount. Take a look at the formula used to calculate simple interest as well as an example that uses this formula.

SIMPLE INTEREST FORMULA

Let I = amount of interest charged or earned

P = original amount borrowed, or principal

r = annual interest rate

t = time in years

Formula: $I = P \times r \times t$ or $I = Prt$

Titus borrowed $4,550 for 5 years at an annual interest rate of 8.5%. How much money must he repay in all?

ANSWER: Use the formula I = Prt to find the amount of interest that will be paid, and then add that amount to the principal amount.

$P = \$4{,}550$ $r = 8.5\%$ $t = 5$

$I = \$4{,}550 \times 0.085 \times 5 = \$1{,}933.75$ Total $= \$4{,}550 + \$1{,}933.75 = \$6{,}483.75$

SKILLS TIP

When setting up a proportion, the like units are placed directly across from each other (numerator and numerator or denominator and denominator). It may be helpful to write a verbal model that identifies the units that are being compared. For example, if you are comparing the number of red marbles to blue marbles, the verbal model might look like:

$\frac{\text{red}}{\text{blue}} = \frac{\text{red}}{\text{blue}}$

SKILLS TIP

If you know three out of the four pieces of information that are in the simple interest formula, then you can use the simple interest formula to find the piece of information that is missing.

Percent Problems

Percent problems can be solved by using the following formula.

PERCENT FORMULA

Amount of Change = Percent × Original Amount

The number of members of a club increased by 5%. There were originally 120 members. How many members are now in the club?

ANSWER: Amount of Change = Percent × Original

$A = 0.05 \times 120 = 6$ new members

$120 + 6 = 126$ total members

Markup and Markdown

A **markup** is simply an increase in the price of an item, while a **markdown** is a discount or a decrease in the price of an item.

A store normally sells a shirt for $22.50. The shirt is discounted by 12%. What is the sale price of the shirt?

ANSWER: Method 1:

Discount = Percent × Original

$D = 0.12 \times 22.50 = \$2.70$

$\$22.50 - \$2.70 = \$19.80$

Method 2:

100% − 12% = 88% of the original price

Sale Price = 0.88 × $22.50 = $19.80

COMMISSION FORMULA

Amount of Commission = Percent × Total Sales

Commission

In some lines of work, a worker receives all or some of his or her pay from **commission**, a percentage of the total amount of the sales he or she made.

Laura sold a home for $142,000. If she receives a 3% commission from the sale, then how much is her commission?

ANSWER: Amount of Commission = Percent × Total Sales

$C = 0.03 \times \$142{,}000$

$C = \$4{,}260$

Laura's commission is $4,260.

SKILLS TIP

When you are using the simple interest formula or the percent formula, remember to change the percent to a decimal by moving the decimal two places to the left.

Lesson Practice

Complete the activities below to check your understanding of the lesson content.

1. Josie's total bill at a restaurant is $17.85. She wants to leave her waiter a 15% tip. About how much money should she leave for the tip?

   A $1.00
   B $2.00
   C $2.75
   D $3.50
   E $4.00

2. A store pays $600 for a washing machine. They sell the washing machine with a 60% markup. What will be the price of the washing machine in the store?

   A $360
   B $420
   C $600
   D $780
   E $960

3. A bank lent $15,500 to Zoe for the purchase of a new car. The loan is for 6 years with an annual interest rate of 6.5%. What is the total amount of interest that Zoe will pay for this loan?

   A $5,500
   B $6,045
   C $7,145
   D $21,545
   E $60,450

4. Daiki buys a pair of jeans that originally cost $44.95. They are on sale for 35% off the original price. How much did Daiki pay for the jeans?

   A $15.73
   B $26.97
   C $29.22
   D $31.47
   E $44.60

5. A survey is given to a group of students. It asks each student to name his or her favorite type of pet. The results are shown in the table below. What percent of the students who took the survey named a dog as their favorite pet?

Favorite Pet	Number
Cat	11
Dog	22
Fish	7
Hamster	10

   A 22%
   B 36%
   C 44%
   D 54%
   E 100%

6. Caleb earns 6% commission on all of his sales. His total commission for the past month was $4,800. What was the amount of his total sales?

   A $288

   B $8,000

   C $75,200

   D $80,000

   E $84,800

7. Over the course of a four-year loan, Everitt paid $1,540 in simple interest. If the annual interest rate was 5%, then how much money did he borrow?

   A $30,800

   B $7,700

   C $5,500

   D $308

   E $77

8. Last month, 45% of the members of a club voted in favor of building an addition to their clubhouse. If 81 members voted in favor of the addition, then how many total members are in the club?

   A 36

   B 99

   C 126

   D 162

   E 180

See page 34 for answers and help.

Answer the questions based on the content covered in this unit.

1. In the last election, 1,200 people in the precinct voted. The ratio of men to women was 3 to 5. How many of the voters were women?
   - **A** 450
   - **B** 480
   - **C** 720
   - **D** 750
   - **E** 850

2. Andy has a piece of lumber that is $7\frac{5}{8}$ feet long. He needs to cut a board that is $5\frac{3}{4}$ feet long. What is the length of the piece of the board that he needs to cut off?
   - **A** $1\frac{1}{4}$ feet
   - **B** $1\frac{3}{4}$ feet
   - **C** $1\frac{7}{8}$ feet
   - **D** $2\frac{7}{8}$ feet
   - **E** $3\frac{1}{8}$ feet

3. Which of the following statements is false?
   - **A** All real numbers are rational.
   - **B** All natural numbers are also whole numbers.
   - **C** A non-terminating, non-repeating decimal number is irrational.
   - **D** All rational numbers are real.
   - **E** The number 0 is a whole number.

4. There are 65 students out of 125 students who play on an athletic team. What is the ratio of students who do not play on an athletic team to all of the students?
   - **A** 12:25
   - **B** 13:25
   - **C** 1:2
   - **D** 13:38
   - **E** 25:38

5. At 10 p.m., the temperature was 11°F. At 5 a.m., the temperature was −10°F. What was the average drop in temperature per hour?
   - **A** 1.5°F
   - **B** 2°F
   - **C** 2.5°F
   - **D** 3°F
   - **E** 3.5°F

6. Salim earns an 8% commission for each car that he sells, plus a $100 base salary each week. How much does he earn for a week in which he sells $78,000 worth of cars?
   - **A** $5,440
   - **B** $6,240
   - **C** $6,340
   - **D** $7,800
   - **E** $7,900

7. A store marks up the original cost of its items by 30%. Which expression can be used to determine the price that the store charges for an item that was originally bought for $15?
   - **A** $15 - 0.3(15)$
   - **B** $15 + 0.3(15)$
   - **C** $\frac{15}{0.3}$
   - **D** $0.3(15)$
   - **E** $\frac{15}{0.3(15)}$

8. Hershel offered to pay for the lunches of his two friends plus his own lunch. Each lunch cost $6.46. He paid with a $20 bill. How much change did he receive?
   - **A** $0.62
   - **B** $7.08
   - **C** $13.54
   - **D** $16.68
   - **E** $19.38

9. Kaitlyn played in a four-day golf tournament. Her score each day compared to par for the course is shown in the table. What is Kaitlyn's score at the end of the tournament?

Day	Score
Thursday	−2
Friday	3
Saturday	−1
Sunday	−4

A 10
B 4
C −2
D −4
E −10

10. Nora borrowed $10,275 at a rate of 6% annual simple interest for a time period of 5 years. At the end of the term of the loan, how much money will she have repaid?
A $616.50
B $3,082.50
C $7,192.50
D $10,891.50
E $13,357.50

11. Shelly needs to raise $161.50 to go on a trip. She earns $4.25 for each pizza she sells. How many pizzas does she need to sell in order to raise enough money to go on the trip?
A 48
B 42
C 38
D 32
E 28

12. Judy is comparing prices of fruit at various stores. The table shows the advertised price per pound for apples at 5 different stores. Which store has the best price?

Store	Pounds	Price
A	3	$4.20
B	2	$2.86
C	5	$6.50
D	10	$12.80
E	4	$5.00

A Store A
B Store B
C Store C
D Store D
E Store E

Questions 13 and 14 refer to the information below.

The following table shows the number of free throws that were attempted and made per game by some of the season-scoring leaders during the 2014–15 NBA season.

2014–15 NBA Leaders		
Player	FT Attempted	FT Made
Russell Westbrook	9.8	8.1
James Harden	10.2	8.8
LeBron James	7.7	5.4
Anthony Davis	6.8	5.5
Stephen Curry	4.2	3.9

13. According to the table, which player had the highest percentage of free throws made per game?
A Russell Westbrook
B James Harden
C LeBron James
D Anthony Davis
E Stephen Curry

14. According to the table, which player had the lowest percentage of free throws made per game?
    - **A** Russell Westbrook
    - **B** James Harden
    - **C** LeBron James
    - **D** Anthony Davis
    - **E** Stephen Curry

15. Pedro has three-fourths of a pizza left over. The next day, he ate half of the leftovers for lunch. What fractional part of the pizza is left?
    - **A** $\frac{3}{2}$
    - **B** $\frac{3}{4}$
    - **C** $\frac{1}{2}$
    - **D** $\frac{3}{8}$
    - **E** $\frac{1}{4}$

16. A scale model of the One World Trade Center building is made to an approximate scale of 1 foot to 325 feet. If the actual height of the One World Trade Center building is 1,776 feet, then about how tall is the scale model?
    - **A** 4.5 feet
    - **B** 5.0 feet
    - **C** 5.5 feet
    - **D** 6.0 feet
    - **E** 6.5 feet

17. Which number is 506,273.9148 rounded to the nearest hundredth?
    - **A** 506,300
    - **B** 506,273.91
    - **C** 506,273.92
    - **D** 506,273.915
    - **E** 506,000

18. In a school, there are 5 boys and 1 teacher for every 4 girls. What is the ratio of teachers to students?
    - **A** 1:9
    - **B** 1:4
    - **C** 1:5
    - **D** 4:5
    - **E** 1:10

Questions 19 and 20 refer to the information below.

A bank has a discounted interest rate for auto loans for a limited time.

For a limited time only ...

Finance your auto purchase for ONLY 2% annual interest for 60 months!

Reduced* for a limited time only!

19. Shanika purchases a car that costs $11,475 during the advertised special. How much interest will she pay over the 60 months?
    - **A** $229.50
    - **B** $1,147.50
    - **C** $3,442.50
    - **D** $12,622.50
    - **E** $13,770

20. If the normal interest rate is 6% annual interest, how much interest will Shanika save by buying the car during this limited time offer?
    - **A** $229.50
    - **B** $1,147.50
    - **C** $2,295
    - **D** $3,442.50
    - **E** $14,917.50

See page 34 for answers.

Lesson 1

1. **E.** The square root of 27 is a non-terminating, non-repeating decimal.

2. **B.** The digit in the ones place of 31,092.4758 is 2, which is less than 5, so the digit in the tens place will remain the same—9.

3. **C.** Choose the range which included all values that will round to 6,000 when rounded to the nearest thousand.

4. **C.** The digit in the tenths place of the number 895.62 is 6, which is more than 5, so the digit in the ones place increases by 1 from 5 to 6.

5. **B.** Integers are composed of both positive and negative numbers, but whole numbers are composed of only positive numbers.

6. **C.** Ed's score of −11 is the lowest score. Chris's score of −8 is greater than Ed's score of −11. Dequan's score of −6 is greater than Chris's score of −8. Payton's score of −2 is greater than Dequan's score of −6.

7. **A.** Point *A* is farther to the left on the number line than Point *D*, so its value is less.

8. **C.** The square root of 40 is a non-terminating, non-repeating decimal.

Lesson 2

1. **C.** Multiply the number of friends, 24, by three-fourths, $24 \times \frac{3}{4} = 18$.

2. **B.** Subtract the value of both checks from the balance, and then add the value of the deposit to the balance.
   $\$327.58 - \$78.55 = \$249.03$
   $\$249.03 - \$129.14 = \$119.89$
   $\$119.89 + \$178.32 = \$298.21$

3. **C.** First, calculate the cost of making the benches. $\$30 \times 25 = \$750$. Next, find the number of benches that he needs to sell in order to cover his costs by dividing his costs, \$750, by his selling price, \$75.
   $\$750 \div \$75 = 10$.

4. **D.** Divide the number of packages of crackers that are fed to the children each weekday by 2 or multiply by $\frac{1}{2}$.
   $4\frac{1}{2} \times \frac{1}{2} = \frac{9}{2} \times \frac{1}{2} = \frac{9}{4} = 2\frac{1}{4}$

5. **B.** Divide the loan amount by the amount that is paid back each week to find the number of weeks that it will take to pay off the loan.
   $\$33.75 \div \$2.50 = 13.5$
   Since the answer is a decimal that is more than 13, round up to 14.

6. **D.** First, figure out how much money she has left over each month after paying for her living expenses by subtracting. $\$3,500 - \$2,650 = \$850$. Next, divide the amount of the car loan by the amount that she has left over each month in order to find the number of months it will take her to pay off the car loan.
   $\$9,000 \div \$850 \approx 10.6$
   Since the answer is a decimal that is more than 10, round up to 11.

7. **A.** First, calculate the difference in pay per month for a teacher and for a teacher's aide by subtracting.
   $\$3,500 - \$2,400 = \$1,100$
   Next, multiply the difference by 12 to calculate the difference in pay for an entire year.
   $\$1,100 \times 12 = \$13,200$

Lesson 3

1. **E.** The expression can be found by first dividing the total number of pounds of meat, 375 pounds, by the number of pounds of meat that are in one box, 15 pounds, to find the total number of boxes of meat that are needed. Next, multiply the number of boxes of meat that are needed by the cost per box of meat, \$40.

2. **D.** First, find the price per box by dividing the price of the 3 boxes, \$5, by the number of boxes, 3.
   $\$5 \div 3 = \frac{5}{3}$ or $1\frac{2}{3}$
   Next, multiply the price per box by the number of boxes, 5.
   $\frac{5}{3} \times 5 = \frac{25}{3} = 8\frac{1}{3}$ or \$8.33

3. **E.** For each answer choice, divide the price by the number of ounces. The answer choice with the lowest price per ounce is the correct answer.

Lesson 4

1. **C.** To find the tip amount, multiply the bill total, \$17.85, by the tip percentage, 15%, in decimal form, 0.15.
   $\$17.85 \times 0.15 = \$2.68$
   Since the problem is asking for an estimate, look for the answer choice that is closest to \$2.68.

2. **E.** First, convert 60% into a decimal, 0.60, and then multiply it by the amount that the store paid for the washing machine, \$600. $\$600 \times 0.60 = \$360$
   Next, add the result to the price that the store paid for the washing machine, \$600. $\$600 + \$360 = \$960$.

3. **B.** Use the simple interest formula, $I = Prt$. First, make sure to convert the percentage, 6.5%, into a decimal, 0.065. Next, plug in all of the known values and then multiply.
   $I = \$15{,}500 \times 0.065 \times 6 = \$6{,}045$

4. **C.** First, multiply the cost of the jeans, \$44.95, by the decimal form of the sale percentage, 0.35, in order to find the amount of money that will be saved.
   $\$44.95 \times 0.35 = \$15.73$
   Next, subtract the amount of money saved from the original price of the jeans in order to find the sale price.
   $\$44.95 - \$15.73 = \$29.22$

5. **C.** First, find the number of students who selected a dog, 22. Next, add up all of the results in order to find the total number of students who took the survey. $11 + 22 + 27 + 10 = 50$
   Finally, divide the number of students who selected a dog by the total number of students who took the survey.
   $22 \div 50 = 0.44$ or 44%.

6. **D.** Use the commission formula, Amount of Commission = Percent × Total Sales. First, convert the commission percentage, 6%, to a decimal, 0.06. Next, plug all of the known values into the formula.
   $\$4{,}800 = 0.06 \times \text{Total Sales}$
   Finally, solve for the total amount of his sales by dividing his commission amount by his commission percentage in decimal form.
   $\$4{,}800 \div 0.06 = \$80{,}000$

7. **B.** Use the interest formula. $I = Prt$.
   First, convert the percentage, 5%, into a decimal, 0.05. Next, plug all of the known values into the formula.
   $\$1{,}540 = P \times 0.05 \times 4$
   Next, multiply the rate in decimal form by the time.
   $0.05 \times 4 = 0.2$
   Finally, divide the total amount of interest by the previous result.
   $\$1{,}540 \div 0.2 = \$7{,}700$

8. **E.** Use the percent formula. Amount of Change = Percent × Original Amount. First, convert the percentage, 45%, into a decimal, 0.45. Next, plug all of the known values into the formula.
   $81 = 0.45 \times \text{Original Amount}$
   Finally, divide the number of members who voted in favor of building the clubhouse addition by the rate in decimal form. $81 \div 0.45 = 180$

Unit Test

1. D.
2. C.
3. A.
4. A.
5. D.
6. C.
7. B.
8. A.
9. D.
10. E.
11. C.
12. E.
13. E.
14. C.
15. D.
16. C.
17. B.
18. A.
19. B.
20. C.

Unit Glossary

- **commission** – an amount of money that is paid to a worker based on a percentage of his or her sales
- **integer** – any number that is not represented by a fraction or decimal
- **irrational numbers** – numbers that are non-terminating, non-repeating decimals
- **markdown** – a discount or decrease in the price of an item
- **markup** – an increase in the price of an item
- **natural numbers** – numbers that are used for counting, such as 1, 2, 3, ...
- **percent** – a portion or part that is based out of 100
- **place value** – the value of each digit found in a number
- **proportion** – two ratios that are equal to each other in value
- **radical number** – a number that has a $\sqrt{\ }$ sign; also called a square root
- **ratio** – a comparison of two or more quantities that have different units
- **rational number** – any number that can be written as a fraction of two integers
- **real number** – any number that can be represented on a number line
- **simple interest** – interest that is paid to the lender and is computed yearly based on the original loan amount
- **unit rate** – comparison of an amount of one unit to a single amount of another unit, such as miles per gallon
- **whole numbers** – all of the natural numbers or counting numbers including 0

Study More!

UNIT 1

Understanding the Real Number System

- Understand into which subsets of real numbers various different types of numbers fit.
- Read numbers correctly using place value.
- Place numbers on a number line and compare two or more numbers by placing them on a number line.
- Evaluate square roots and radical expressions.

Performing Operations on Real Numbers

- Add, subtract, multiply, and divide mixed numbers.
- Understand how to add, subtract, multiply, and divide two or more positive and negative numbers.

Describing a Ratio between Two Quantities

- Understand and use unit rates that have units other than money.

Using Ratios and Proportions to Solve Problems

- Use the percent change formula.
- Solve problems that involve a percent greater than 100.
- Solve percent problems that involve "increased by" or "decreased by."
- Solve problems that involve profit.
- Use proportions to solve percent problems.
- Solve problems that involve compound interest.

PERCENT CHANGE FORMULA

$$\text{percent of change} = \frac{\text{amount of change}}{\text{original amount}}$$

UNIT 2

Operations and Algebraic Thinking

Equations can be used to solve real-world problems that have missing information. Use variables to represent the unknown information. Consider the following scenario:

> Jermaine promotes his plumbing business on a social media site as shown in the ad above. Mrs. Fernandez hires Jermaine to fix her plumbing. When the work is complete, she writes him a check for \$335 for work performed during regular business hours. Write and solve an equation that can be used to determine the number of hours that Jermaine worked at Mrs. Fernandez's home.

Answer: $65 + 45x = 335$; 6 hours

KEY WORDS

- algebraic expression
- Associative Property of Addition
- Associative Property of Multiplication
- Commutative Property of Addition
- Commutative Property of Multiplication
- constant
- Distributive Property
- equation
- one-step equations
- two-step equations

Commutative Properties of Addition and Multiplication

The **Commutative Property of Addition** and the **Commutative Property of Multiplication** both say that the order in which you add or multiply two real numbers does not change the sum or product. So, for all real numbers:

Commutative Property of Addition	Commutative Property of Multiplication
$a + b = b + a$	$ab = ba$
Example: $3 + 2 = 2 + 3$	Example: $3 \cdot 2 = 2 \cdot 3$

The Commutative Properties can be used to help simplify arithmetic and algebraic expressions. Reorder the numbers in the following expressions and then simplify.

$16 + 9 + 4 + 11$

ANSWER:

$16 + 4 + 9 + 11$

$20 + 20$

40

$4x \cdot 5y$

ANSWER:

$4 \cdot 5 \cdot x \cdot y$

$20xy$

SKILLS TIP

The Commutative Property allows the order of the numbers to be changed to make the addition or multiplication easier to compute. Simply rearrange the numbers so that they are simpler to add or multiply, then perform the addition or multiplication.

Associative Properties of Addition and Multiplication

The **Associative Property of Addition** and the **Associative Property of Multiplication** both say that when you add or multiply any three real numbers, the grouping (or association) does not matter. The grouping symbols can be rearranged without having an effect on the result. So, for all real numbers:

Associative Property of Addition	Associative Property of Multiplication
$(a + b) + c = a + (b + c)$	$(ab)c = a(bc)$
Example: $(3 + 2) + 6 = 3 + (2 + 6)$	Example: $(3 \cdot 2)4 = 3(2 \cdot 4)$

The Associative Properties can also be used to help simplify arithmetic and algebraic expressions. Rearrange the parentheses in the following expressions and then simplify.

$(64 + 98) + 2$

ANSWER:

$64 + (98 + 2)$

$64 + 100$

164

$\frac{1}{3}(9x)$

ANSWER:

$(\frac{1}{3} \cdot 9)x$

$3x$

Using the Properties of Addition and Multiplication

Distributive Property

The **Distributive Property** is used to simplify expressions wherein a number is multiplied by an addition or subtraction expression. The addition or subtraction expression is enclosed in parentheses. The Distributive Property simplifies the process of multiplying.

Distributive Property – Addition	Distributive Property – Subtraction
For all real numbers a, b, and c: $a(b + c) = ab + ac$ and $(b + c)a = ba + ca$	For all real numbers a, b, and c: $a(b - c) = ab - ac$ and $(b - c)a = ba - ca$
Example: $2(x + 4) = 2 \cdot x + 2 \cdot 4 = 2x + 8$	Example: $2(x - 3) = 2 \cdot x - 2 \cdot 3 = 2x - 6$

SKILLS TIP

Parentheses in an expression or equation provide a clue to check whether the Distributive Property can be used to simplify it.

Use the Distributive Property so that mental math can be used to solve these problems.

a. $6 \cdot 52$

ANSWER:

a. $6 \cdot 52 = 6(50 + 2)$
$= (6 \cdot 50) + (6 \cdot 2)$
$= 300 + 12$
$= 312$

b. $8(6.5)$

ANSWER:

b. $8(6.5) = 8(6 + 0.5)$
$= (8 \cdot 6) + (8 \cdot 0.5)$
$= 48 + 4$
$= 52$

Lesson Practice

KEY POINT!

The Commutative Property of Addition allows for the order of the numbers to be changed without affecting the sum, and the Associative Property of Addition allows for the placement of the parentheses to be moved without affecting the sum.

Complete the activities below to check your understanding of the lesson content.

1. Which of the following statements demonstrates the Commutative Property of Addition?

   A $9 \cdot 8 = 8 \cdot 9$
   B $9 + 8 = 8 + 9$
   C $7 + 6 = 8 + 5$
   D $3 \cdot 4 = 2 \cdot 6$
   E $4 + 6 = 12 - 2$

2. Which of the following statements demonstrates how the Distributive Property can be used to find the product of $4 \cdot 76$?

   A $76 \cdot 4$
   B $(4 + 70) \cdot (4 + 6)$
   C $4(7 + 6) = 4 \cdot 7 + 4 \cdot 6$
   D $4(70 + 6) = 4 \cdot 70 + 4 \cdot 6$
   E $4(70 + 6) = (4 + 70)6$

3. A movie theater sells bags of popcorn for $7 and large soft drinks for $5. On Saturday, the theater sold exactly 3 times the amount of popcorn and soft drinks as they did on Friday. Which expression can be used to determine the amount of money that the theater received for its sales of popcorn and soft drinks on Friday and Saturday?
   - **A** $7x + 5y$
   - **B** $3 \cdot 7x + 5y$
   - **C** $2(7x + 5y)$
   - **D** $3(7x + 5y)$
   - **E** $3(7x + 5y) + 7x + 5y$

4. The perimeter of a rectangle can be found by using the formula $P = 2l + 2w$, where l is the length of the rectangle and w is the width of the rectangle. Which expression can be used to find the perimeter of the rectangle that is shown below?

   $x + 3$ ft

   x ft

   - **A** $x + x + 3$
   - **B** $x + 2(x + 3)$
   - **C** $2x + 2x + 3$
   - **D** $2x + 2(x + 3)$
   - **E** $x(x + 3)$

5. Samir has A quarters in one jar and B quarters in a different jar. The total value of the change is $26.25. Which equation can be used to represent the number of quarters in each jar?
   - **A** $\$0.25(A + B) = \$26.25$
   - **B** $A + B = \$26.25$
   - **C** $A + B + \$0.25 = \$26.25$
   - **D** $\$0.25(A \cdot B) = \$26.25$
   - **E** $\$0.25(A - B) = \$26.25$

6. Which of the following statements is false?
   - **A** The Commutative Property of Addition indicates that the order of the numbers does not change the sum when adding.
   - **B** The Distributive Property is often used to simplify expressions with parentheses.
   - **C** The Distributive Property cannot be used with decimals.
   - **D** The Associative Property of Multiplication involves changing the placement of the parentheses.
   - **E** The Commutative Properties of Addition and Multiplication always work when used correctly.

7. A class trip to Washington, D.C., will cost a total of $18,000. The class has saved up $11,500 through fundraising. There are 15 students who will go on the trip. Which expression represents the amount that each student will have to pay in order to go on the trip?
   - **A** $\$18{,}000 + \$11{,}500 \div 15$
   - **B** $\$18{,}000 - \$11{,}500 \div 15$
   - **C** $(\$18{,}000 - \$11{,}500) \div 15$
   - **D** $(\$18{,}000 + \$11{,}500) \div 15$
   - **E** $(\$18{,}000 - \$11{,}500) \times 15$

TEST STRATEGY

Read each problem carefully. Then read all of the answer choices carefully. Some answer choices might look correct but do not answer the question that is asked. Make sure you are answering the question that is asked.

TEST STRATEGY

For some problems, it is better to set up an equation that can be used to solve the problem before you look at the answer choices.

KEY POINT!

Review the properties that were discussed in the lesson. The Commutative Property refers to the order of the numbers that are being added or multiplied. The Associative Property refers to the replacement of the parentheses around a different group of numbers when adding or multiplying. The Distributive Property is used to multiply a number by an expression in parentheses.

Lesson Practice

Questions 8 and 9 refer to the information below.

Watertown has commissioned an artist to create a tile mosaic. The mosaic will be composed of same-sized, different-colored square tiles framed and have a red rectangular frame around it. The drawing below shows the entire mosaic along with its dimensions.

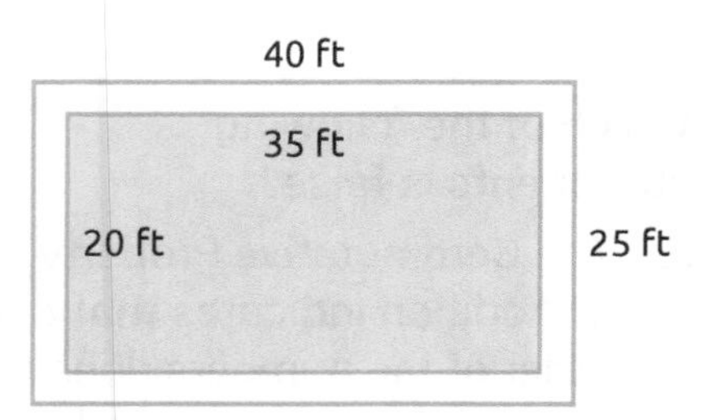

8. Which expression can be used to determine the area of the frame around the mosaic?

   A $(40 \times 25) \div 35 \times 20$

   B $(40 + 25) - (35 + 20)$

   C $(40 \times 25) + (35 \times 20)$

   D $(40 \times 25) - (35 \times 20)$

   E $(40 \times 25) \div (35 \times 20)$

9. Each square tile that is used for the mosaic has a side length of 5 feet. Which of the following expressions can be used to determine the number of square tiles that the artist will need to create the mosaic?

   A $(5 \times 5) \div (35 \times 20)$

   B $(35 \times 20) \div (5 \times 5)$

   C $(40 \times 25) \div (5 \times 5)$

   D $(40 \times 25) \div (35 \times 20)$

   E $(35 \times 20) - (5 \times 5)$

See page 49 for answers and help.

10. Which of the following expressions represents *a* hours, *b* minutes, and *c* seconds in terms of total hours?

    A $a + 60b + 3{,}600c$

    B $60a + 120b + 360c$

    C $\frac{a}{60} + b + 60c$

    D $a + \frac{b}{60} + \frac{c}{3600}$

    E $a + 60b + 360c$

11. Which of the following demonstrates the Associative Property of Addition?

    A $3 + (17 + 11) = (3 + 17) + 11$

    B $\frac{1}{4} \times 7 \times 8 = \frac{1}{4} \times 8 \times 7$

    C $\frac{1}{4} \times (16 \times 11) = (\frac{1}{4} \times 16) \times 11$

    D $12 + 17 + 8 = 12 + 8 + 17$

    E $4(x + 3) = 4x + 12$

12. Which of the following demonstrates the Commutative Property of Multiplication?

    A $6 + (14 + 9) = (6 + 14) + 9$

    B $\frac{1}{7} \times 9 \times 35 = \frac{1}{7} \times 35 \times 9$

    C $\frac{1}{5} \times (15 \times 7) = (\frac{1}{5} \times 15) \times 7$

    D $(23 + 9) + 7 = 23 + (9 + 7)$

    E $5(x - 7) = 5x - 35$

KEY POINT!

Remember, the Distributive Property can be used to rewrite the multiplication of a whole number by a decimal number. Split the decimal number into the sum of a whole number part and a decimal part before multiplying.

Word Sentences and Equations

Word phrases can be translated into algebraic expressions and equations. When solving a word problem, the first step is to take the situation that is described in words and rewrite it as an expression or equation. To do this, you need to be able to identify the key words and understand how to translate them into a number sentence or equation. Here are some key words that represent each operation.

Addition	Subtraction	Multiplication	Division	Equals
add	minus	times	quotient	equals
plus	subtract	product	divide	is
sum	difference	multiply	half	result
more than	less than			
increased by	decreased by			

Write an algebraic expression for each phrase shown. Use x to represent the unknown number.

a. Nine less than a number — **ANSWER:** $x - 9$

b. Two times the sum of a number and four — **ANSWER:** $2(x + 4)$

c. Four times a number is twelve — **ANSWER:** $4x = 12$

One-Step Equation Word Problems

An **algebraic expression** is a number sentence that does not have an equal sign. An expression can be simplified, but not solved. An **equation** has an equal sign and can be solved by isolating the variable. In some cases, it only takes one step to get the variable alone on one side of the equation.

Yesterday, Isabel drove 140 miles at a constant speed of 35 miles per hour. How many hours was Isabel driving? Write and solve an equation.

ANSWER: Use the equation $d = rt$, where d is the distance, r is the rate of speed, and t is the time.

$140 = 35t$ Substitute 140 for d and 35 for r.

$\frac{140}{35} = \frac{35t}{35}$ Divide both sides by 35.

$4 = t$ Simplify.

Isabel drove for 4 hours.

SKILLS TIP

To undo multiplication, divide. If x is being multiplied by 2, divide both sides by 2 to isolate x. Remember, whatever you do to one side of the equation, you must also do to the other side of the equation.

Interpreting and Solving Word Problems

SKILLS TIP

A **constant** is a real number that does not have a variable attached to it.

Two-Step Equation Word Problems

In some equations, it takes two steps to isolate the variable. Begin by moving the constant to the other side of the equation by undoing the addition or subtraction. After the constant is on the other side of the equation, remove the coefficient in order to isolate the variable.

Solve: $3x + 5 = 20$

ANSWER: $3x + 5 = 20$

$-5 = -5$ Subtract 5 from both sides.

$3x = 15$ Simplify.

$\frac{3x}{3} = \frac{15}{3}$ Divide both sides by 3.

$x = 5$ Simplify.

SKILLS TIP

Read the problem carefully. The daily fee is a flat fee. The number of miles driven is what increases the daily cost. Therefore, the number of miles driven is the value that is represented with a variable.

Word problems often require two-step equations, like in the following example.

Lamont rented a car for a one-day business trip. The rental company charges $20 per day plus $0.40 per mile. The total charge for the day was $87.20. How many miles did Lamont drive the car?

ANSWER: Let *x* represent the unknown number of miles that Lamont drove the car. So, the equation must multiply the unknown number of miles, *x*, by $0.40 and then add the daily fee of $20.

$\$0.40x + \$20 = \$87.20$

$-\$20 = -\$20$ Subtract 20 from both sides.

$\$0.40x = \$67.20$ Simplify.

$\frac{\$0.40x}{\$0.40} = \frac{\$67.20}{\$0.40}$ Divide both sides by 0.40.

$x = 168$ Simplify.

Lamont drove the rental car 168 miles.

Complete the activities below to check your understanding of the lesson content.

1. In the 2014 NFL regular season, the Denver Broncos won 4 times as many games as the Oakland Raiders. If Denver won 12 games, then which of the following equations can be used to find the number of games that Oakland won?

   A $12 - w = 4$, where w is the number of Oakland wins

   B $4 + w = 12$, where w is the number of Oakland wins

   C $12w = 4$, where w is the number of Oakland wins

   D $4w = 12$, where w is the number of Denver wins

   E $4w = 12$, where w is the number of Oakland wins

Questions 2 and 3 refer to the information below.

Tina is three years older than twice Darnell's age. Tina is 39 years old.

2. Which of the following equations can be used to determine Darnell's age?

   A $2x + 3 = 39$

   B $x + 3 = 39$

   C $2x + 39 = 3$

   D $2x - 3 = 39$

   E $x + 39 = 3$

3. How old is Darnell?

   A 3

   B 9

   C 13

   D 18

   E 36

4. Three less than nine times a number is 42. Which of the following equations represents this word sentence?

   A $3 - 9n = 42$

   B $3 + 9n = 42$

   C $9n - 3 = 42$

   D $3 \cdot 9 - n = 42$

   E $n - 3 = 42$

Questions 5 and 6 refer to the information below.

The following table shows the prices for several fees that are associated with going to a fair.

Fees	Cost
Adult Admission	$8.50
Child Admission	$4.50
Parking	$5
Ride Pass	$7

5. A group of senior citizens is planning to attend the fair and not go on any of the rides. They take two vehicles, and the total cost is $86.50. If there are no children in the group and no one buys a ride pass, then which of the following equations can be used to determine the number of senior citizens who are going to the fair?

   A $\$8.50a = \$86.50$

   B $\$8.50a + \$5 = \$86.50$

   C $\$8.50a + \$10 = \$86.50$

   D $\$8.50a + \$7 = \$86.50$

   E $\$8.50a + \$14 = \$86.50$

KEY POINT!

Identify key words in the problem that indicate operations such as "times" (multiplied by) and "is" (equals) to help you set up the equation.

Lesson Practice

TEST STRATEGY

Read the problem carefully to make sure that you use only the information that is pertinent to answering the question that is asked. Sometimes, problems include information that does not need to be used to find the solution.

KEY POINT!

The word "sum" indicates addition.

TEST STRATEGY

Reread the problem and check to make sure that your answer makes sense.

6. How many senior citizens are in the group that is going to the fair?
   - A 4
   - B 9
   - C 10
   - D 12
   - E 13

7. The observatory on the 102nd floor of the Empire State Building is 1,250 feet off the ground. The antenna of the building starts on the 102nd floor, and its tip is 1,454 feet off the ground. Which of the following equations can be used to determine the height of the antenna?
   - A $x - 1{,}454 = 1{,}250$
   - B $x - 1{,}250 = 1{,}454$
   - C $1{,}250 + x = 1{,}454$
   - D $1{,}250 - x = 1{,}454$
   - E $1{,}250x = 1{,}454$

8. Maya buys 14 passes to ride a bus to work. She pays \$35. How much is one trip to work on the bus?
   - A \$2.50
   - B \$3.50
   - C \$11
   - D \$35
   - E \$49

9. The sum of three times a number and 15 is 45. What is the number?
   - A 5
   - B 10
   - C 15
   - D 30
   - E 60

See page 49 for answers and help.

Questions 10 and 11 refer to the information below.

In the school election, 60 students, or $\frac{3}{4}$ of the student body, voted for Tanikka for class president.

10. Which of the following equations can be used to determine the number of students in the student body?
    - A $x - \frac{3}{4} = 60$
    - B $\frac{3}{4} - x = 60$
    - C $\frac{3}{4} + x = 60$
    - D $\frac{3}{4}x = 60$
    - E $\frac{3}{4}(60) = x$

11. How many students are there in the student body?
    - A 15
    - B 45
    - C 80
    - D 90
    - E 100

12. A bookstore offers 20% off any regular-priced item. Daria paid \$15.96 for a book. Which of the following equations can be used to determine the original price of the book?
    - A $0.2x = \$15.96$
    - B $0.8x = \$15.96$
    - C $1.2x = \$15.96$
    - D $x - 0.2 = \$15.96$
    - E $x + 0.2 = \$15.96$

Answer the questions based on the content covered in this unit.

1. A charity is attempting to raise $50,000. The annual fundraising drive has already raised $37,500. There are 5 businesses that have agreed to split the remaining amount that is needed to reach the goal. Which expression can be used to determine the amount of money that each business will have to donate to the charity?

   A $50,000 + $37,500 ÷ 5
   B $50,000 − $37,500 ÷ 5
   C ($50,000 + $37,500) ÷ 5
   D ($50,000 − $37,500) ÷ 5
   E ($50,000 − $37,500) × 5

2. Which of the following demonstrates the Associative Property of Multiplication?

   A $(5 + 2) + 4 = 5 + (2 + 4)$
   B $(5 \times 2)4 = 5(2 \times 4)$
   C $5 + 2 + 4 = 4 + 5 + 2$
   D $5 \times 2 \times 4 = 4 \times 5 \times 2$
   E $3(4 + x) = 12 + 3x$

Questions 3 and 4 refer to the information below.

Adaya earns $25 per lawn that she mows in her neighborhood. Last week, she spent $17.75 on gas for her mower.

3. If Adaya made a profit of $307.25 last week, after paying for gas, which of the following equations can be used to determine how many lawns she mowed?

   A $25*x* − $307.25 = $17.75
   B $25*x* + $307.25 = $17.75
   C $25*x* + $17.75 = $307.25
   D $25*x* − $17.75 = $307.25
   E $17.75*x* − $25 = $307.25

4. If Adaya made a profit of $307.25 last week, after paying for gas, then how many lawns did she mow?

   A 8
   B 11
   C 13
   D 15
   E 17

5. Which of the following demonstrates how the Distributive Property can be used to help find the product of 7 and 11.5?

   A $7 + 11 \times 7 + 0.5$
   B $7(11) \times 7(0.5)$
   C $7(11) \div 7(0.5)$
   D $7(11) - 7(0.5)$
   E $7(11) + 7(0.5)$

Questions 6 and 7 refer to the information below.

Andy is building a patio. The interior of the patio is a rectangle that is made up of red same-sized square brick pavers. A rectangular border surrounds the patio and is made up of gray same-sized square brick pavers. The drawing below shows the entire patio along with its dimensions.

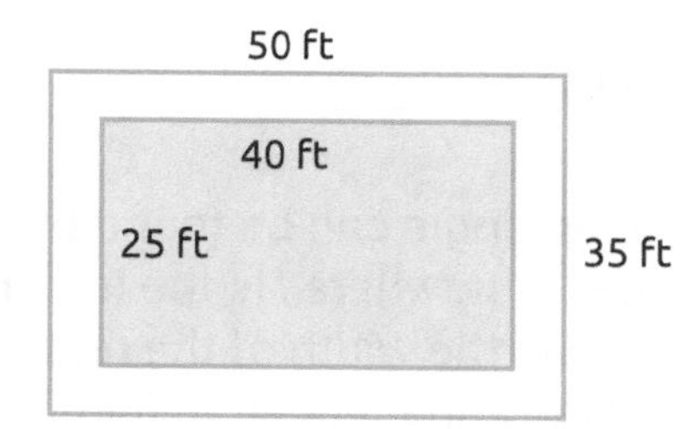

6. Which expression can be used to determine the area of the patio that is covered by the gray pavers?

   A $(50 - 35) \times (40 - 25)$
   B $(50 + 35) - (40 + 25)$
   C $(50 \times 35) - (40 \times 25)$
   D $(50 \times 35) + (40 \times 25)$
   E $(50 \times 35) \div (40 \times 25)$

7. Each red square brick paver has a side length of 1.5 feet. Which of the following expressions can be used to determine the number of square pavers that Andy will need to build the interior of the patio?

   A $(50 \times 35) - (1.5 \times 1.5)$
   B $(50 \times 35) \div (1.5 \times 1.5)$
   C $(40 \times 25) + (1.5 \times 1.5)$
   D $(40 \times 25) \div (1.5 \times 1.5)$
   E $(40 \times 25) - (1.5 \times 1.5)$

Questions 8 and 9 refer to the information below.

The U.S. Pentagon in Washington, D.C., has a total perimeter of 4,605 feet.

8. Which of the following equations can be used to determine the length of each side of the Pentagon (5 sides)?
   A $5x = 4{,}605$
   B $5 + x = 4{,}605$
   C $5 \cdot 4{,}605 = x$
   D $\frac{x}{5} = 4{,}605$
   E $x - 5 = 4{,}605$

9. What is the length of each side of the Pentagon?
   A 5 feet
   B 921 feet
   C 4,600 feet
   D 4,610 feet
   E 23,025 feet

10. The area of a rectangle can be found by using the formula $A = l \times w$, where l is the length of the rectangle and w is the width of the rectangle. Which expression can be used to find the area of the rectangle shown here?

$x - 1$ in

6 in

   A $6x - 6$ in^2
   B $6x - 1$ in^2
   C $12 + 2(x - 1)$ in^2
   D $6x + 2(x - 1)$ in^2
   E $x - 6$ in^2

Questions 11 and 12 refer to the information below.

During the 2014 Baseball season, the Detroit Tigers won 1.25 times more games than they lost. They won 90 games during that season.

11. Which of the following equations can be used to determine how many games the Detroit Tigers lost during the 2014 regular season?
   A $\frac{1.25}{x} = 90$
   B $1.25 + x = 90$
   C $1.25 \cdot 90 = x$
   D $\frac{x}{1.25} = 90$
   E $1.25x = 90$

12. How many games did the Detroit Tigers lose during the 2014 regular season?
   A 18 games
   B 66 games
   C 72 games
   D 90 games
   E 113 games

13. A snack booth at a fair sells corn dogs for \$6 and ice cream cones for \$4. On opening day, the booth sold twice as many corn dogs and ice cream cones as it did on closing day. Which expression can be used to determine the amount of money that the snack booth received for its sales of corn dogs and ice cream cones on both days? Let c be the number of corn dogs sold and i be the number of ice cream cones sold.
   A $2(6c + 4i) + 6c + 4i$
   B $2 \cdot 6c + 4i$
   C $2(6c + 4i)$
   D $2(6c + 4i) + 2(6c + 4i)$
   E $2(6c + 4i) - 6c + 4i$

14. Which of the following statements is false?
    - **A** The Commutative Property of Addition indicates that the order of the numbers does not matter when adding.
    - **B** The Distributive Property is used when there is a number being multiplied by a quantity in parentheses.
    - **C** The Distributive Property can be used when multiplying a whole number by a decimal.
    - **D** The Associative Property can work in certain cases with subtraction.
    - **E** The Commutative Property only works with Addition and Multiplication.

15. Which of the following expressions represents x yards, y feet, and z inches in terms of total yards?
    - **A** $x + y + z$
    - **B** $3(x + y + z)$
    - **C** $x + \frac{y}{3} + \frac{z}{36}$
    - **D** $\frac{1}{3}x + y + 12z$
    - **E** $\frac{1}{3}x + 12y + z$

16. A clothing store in a mall is offering 25% off any regularly priced item. Ethan buys a pair of jeans for $34.50. Which equation can be used to determine the original price of the jeans?
    - **A** $x + 0.25 = \$34.50$
    - **B** $x - 0.25 = \$34.50$
    - **C** $1.25x = \$34.50$
    - **D** $0.75x = \$34.50$
    - **E** $0.25x = \$34.50$

Questions 17 and 18 refer to the information below.

Kalia earns 10% commission on the sales that she makes plus $250 per week as a base salary.

17. Last week, Kalia earned a total of $2,805. Which of the following equations can be used to determine the amount of her total sales for last week?
    - **A** $\frac{0.10}{x} + \$250 = \$2{,}805$
    - **B** $\frac{x}{0.10} + \$250 = \$2{,}805$
    - **C** $1.10 \cdot x + \$250 = \$2{,}805$
    - **D** $0.10x + \$250 = \$2{,}805$
    - **E** $0.10x = \$2{,}805$

18. What was the amount of her total sales for last week?
    - **A** $2,805
    - **B** $25,000
    - **C** $25,550
    - **D** $28,050
    - **E** $30,055

19. Simon used the Distributive Property to rewrite the expression $9(7 + 0.25)$. Which of the following was the original expression?
    - **A** 97×0.25
    - **B** $97 + 0.25$
    - **C** 9×72.5
    - **D** 9×7.25
    - **E** $9 + 7.25$

20. Which of the following demonstrates the Associative Property of Multiplication?
    - **A** $\frac{2}{3} \times 5 \times 12 = \frac{2}{3} \times 12 \times 5$
    - **B** $11 + (9 + 23) = (11 + 9) + 23$
    - **C** $7(2x - 4) = (14x - 28)$
    - **D** $\frac{1}{5}(10 \times 7) = (\frac{1}{5} \times 10)7$
    - **E** $57 + 17 + 3 = 57 + 3 + 17$

21. Which of the following demonstrates the Commutative Property of Addition?
    - **A** $7 + 41 + 13 = 7 + 13 + 41$
    - **B** $-2(5x - 4) = -10x + 8$
    - **C** $\frac{1}{7}(49 \times 6) = (\frac{1}{7} \times 49)6$
    - **D** $26 + (14 + 8) = (26 + 14) + 8$
    - **E** $\frac{1}{9} \times 11 \times 81 = \frac{1}{9} \times 81 \times 11$

22. Which of the following demonstrates the Distributive Property?
    - A $\frac{1}{2}(18 \times 7) = (\frac{1}{2} \times 18)7$
    - B $52 + (18 + 7) = (52 + 18) + 7$
    - C $25 + 13 + 15 = 25 + 15 + 13$
    - D $-3(2x + 6) = (-3)(2x) + (-3)(6)$
    - E $\frac{1}{6} \times 9 \times 36 = \frac{1}{6} \times 36 \times 9$

Questions 23 and 24 refer to the information below.

Camilla purchases enough ride passes so that both of her children can each ride 6 different rides at the fair. She paid $27 for the passes.

23. Which equation can be used to determine the price of an individual ride pass?
    - A $6x = \$27$
    - B $12x = \$27$
    - C $6x + 2 = \$27$
    - D $12x + 2 = \$27$
    - E $12 \cdot \$27 = x$

24. How much did Camilla pay for each individual ride pass?
    - A $2.25
    - B $2.50
    - C $3.25
    - D $3.50
    - E $4.50

25. Write an expression that represents the following word sentence.
    "Six times the difference of twice a number and 5."
    - A $6 + 2x - 5$
    - B $2(6x - 5)$
    - C $6 \cdot 2x - 5$
    - D $6(2x - 5)$
    - E $6(2x + 5)$

Questions 26 and 27 refer to the information below.

The base of the triangle is $x + 1$ feet and the height is x ft.

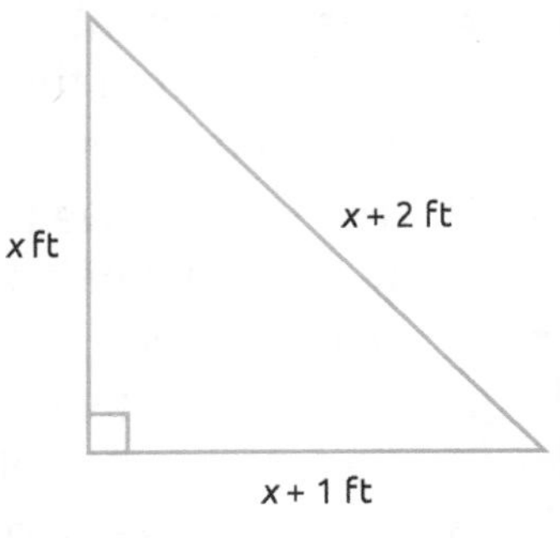

26. The perimeter of a triangle is the sum of the lengths of all its sides. Which of the following expressions represents the perimeter of the triangle?
    - A $x + 3$ feet
    - B $2x + 3$ feet
    - C $3x + 3$ feet
    - D $x(x + 1)$ feet
    - E $x(x + 1)(x + 2)$ feet

27. The formula for the area of a triangle is $\frac{1}{2} \times$ base $\times$ height. Which of the following expressions represents the area of the triangle?
    - A $\frac{1}{2}x(x + 2)$ ft^2
    - B $\frac{1}{2}x(x + 1)$ ft^2
    - C $\frac{1}{2}(x + 1)(x + 2)$ ft^2
    - D $\frac{1}{2}x(x + 1)(x + 2)$ ft^2
    - E $x(x + 1)$ ft^2

28. Fernando charges a flat fee of $4.50 plus $2.00 per mile for his taxi service in the city. When he drove a passenger to the airport, the cab fare was $12.50. How many miles was this trip to the airport?
    - A 2 miles
    - B 3 miles
    - C 4 miles
    - D 5 miles
    - E 6 miles

See page 50 for answers.

Lesson 1

1. **B.** Look for a statement that uses addition and that has a different order of the numbers on each side of the equal sign.

2. **D.** Answer choice D shows how 76 can be rewritten as the sum of the values of its digits so that the quantity can be multiplied by 4 using mental math.

3. **E.** First, write an expression for the amount of money the theater received on Friday, $7x + 5y$. Next, write an expression for the amount of money the theater received on Saturday, $3(7x + 5y)$. Finally, add the expressions, $3(7x + 5y) + 7x + 5y$.

4. **D.**

5. **A.** The value of a quarter is $0.25. The total number of quarters is $A + B$. Multiply the value of a quarter by the total number of quarters, $A + B$, and set the product equal to the total amount of money, $26.25.

6. **C.** The Distributive Property can be used with decimals as well as with fractions.

7. **C.** First, write an expression for the value of the remaining amount of money, $18,000 – $11,500. Since there are 15 students, divide the entire expression, or the difference, for the remaining amount of money by 15. ($18,000 – $11,800) ÷ 15

8. **D.** Subtract the area of the mosaic (the inner rectangle) from the entire area (the larger rectangle)

9. **B.** Divide the area of the mosaic (the inner rectangle) by the area of each square tile.

10. **D.**

11. **A.** Look for an answer choice that uses addition and that has a different placement of parentheses on each side of the equal sign.

12. **B.** Look for an answer choice that uses multiplication and that has a different order of numbers on each side of the equal sign.

Lesson 2

1. **E.** Think, "4 times the number of Oakland wins is equal to the number of Denver wins, 12."

2. **A.** Think, "2 times Darnell's age plus 3 is equal to Tina's age, 39."

3. **D.** Isolate the variable in the equation from question 2, $2x + 3 = 39$. Subtract 3 from both sides of the equation and then divide both sides of the equation by 2.

4. **C.** "Three less" means subtract 3, and "nine times a number" means $9n$. $9n - 3 = 42$

5. **C.** First, find an expression for the total price of admission, $\$8.50a$. Next, figure out the total price for parking 2 vehicles: $2 \times \$5 = 10$ Finally, add all of the costs and set the sum equal to the total cost, $86.50.

6. **B.**

7. **C.** Think, "The height of the Empire State Building, 1,250 ft, plus the height of the antenna, x, is equal to the total height, 1,454 ft."

8. **A.** Think, "The number of passes, 14, times the price per pass, x, is equal to the total cost, $35."

9. **B.** "Three times a number" is represented by $3x$. "The sum" is represented by $3x + 15$. "Is" means equal to. So, $3x + 15 = 45$.

10. **D.** Think, "three-fourths times the number of students in the student body, x, is equal to the number of students who voted for Tanikka."

11. **C.** Isolate the variable in the equation from Question 9, $\frac{3}{4}x = 60$. Multiply both sides of the equation by the reciprocal of $\frac{3}{4}$ which is $\frac{4}{3}$.

12. **B.**

Unit Test

1. D.
2. B.
3. D.
4. C.
5. E.
6. C.
7. D.
8. A.
9. B.
10. A.
11. E.
12. C.
13. A.
14. D.
15. C.
16. D.
17. D.
18. C.
19. D.
20. D.
21. A.
22. D.
23. B.
24. A.
25. D.
26. C.
27. B.
28. C.

Unit Glossary

- **algebraic expression** – a number sentence consisting of constants and variables but no equal sign
- **Associative Property of Addition** – for all real numbers a, b, and c: $(a + b) + c = a + (b + c)$
- **Associative Property of Multiplication** – for all real numbers a, b, and c: $(a \cdot b)c = a(b \cdot c)$
- **Commutative Property of Addition** – for all real numbers a and b: $a + b = b + a$
- **Commutative Property of Multiplication** – for all real numbers a and b: $a \cdot b = b \cdot a$
- **constant** – a term with no variable as a factor; simply a number
- **Distributive Property** – for all real numbers a, b, and c: $a(b - c) = ab - ac$ or $(b - c)a = ba - ca$
- **equation** – a number sentence consisting of constants and variables and which has an equal sign
- **one-step equations** – equations that require one transformation to isolate the variable, or get the variable by itself on one side of the equation
- **two-step equations** – equations that require two transformations to isolate the variable, or get the variable by itself on one side of the equation

UNIT 2

Study More!

Using the Properties of Addition and Multiplication

- Understand the properties of addition and multiplication and how they can be applied to fractions and decimals.
- Understand how the order of operations can be applied to the properties of addition, multiplication, and the Distributive Property.

Interpreting and Solving Word Problems

- Interpret and solve word problems that use fractions and/or decimals.
- Interpret and solve word problems that involve multi-step equations.

UNIT 3

Algebraic Expressions, Equations, and Inequalities

Hidden Cove Beach 375 miles

Algebraic expressions, equations, and inequalities are some of the basic tools used throughout algebra. It is important to have a good understanding of what these are and how to use them to solve problems in algebra and beyond. Consider the following scenario:

> Lester is driving to Hidden Cove Beach for a vacation. The beach is 375 miles away. How long will it take Lester to arrive at the beach if his average speed is 58 miles per hour?
>
> **Answer:** about 6.5 hours

KEY WORDS

- algebraic expression
- Cartesian coordinate plane
- coefficient
- constant
- equation
- extraneous solution
- greatest common factor (GCF)
- inequality
- intercept
- like terms
- monomial
- order of operations
- polynomial
- Product Property of Square Roots
- quadratic equation
- radical
- radical equation
- radicand
- rational expression
- scientific notation
- slope
- slope-intercept form
- system of equations
- variable
- *x*-intercept
- *y*-intercept
- Zero-Product Property

Variables, Constants, and Expressions

Working with numbers is the basis of arithmetic. In algebra, letters such as *x* or *y* (called **variables**) are used to represent unknown values or numbers. A **constant** is a fixed number, such as 7. An **algebraic expression** is a combination of variables and/or constants, such as $x + 8$. Algebraic expressions can be used to represent real-world scenarios.

Janika is 3 years older than Sienna. Write an algebraic expression that represents Janika's age in relation to Sienna's age.

ANSWER: Let *s* represent the age of Sienna. So, Janika is $s + 3$ years old.

SKILLS TIP

When representing a real-world scenario with an algebraic expression or equation, begin by identifying the unknown values, and then assign a variable to each unknown value.

Order of Operations

It's necessary to understand the **order of operations** in order to simplify an expression. The order of operations states the order of the steps that you must follow any time an expression is being simplified. The order of operations can be remembered by PEMDAS:

P **Parentheses** – Simplify anything in parentheses first.

E **Exponents** – Simplify any exponents next.

M **Multiplication** – Multiply and divide working from left to right.

D **Division**

A **Addition** – Add and subtract working from left to right.

S **Subtraction**

Follow the order of operations to simplify $5 - 7 \cdot 4 + 2(4 + 6)$.

ANSWER:

$$\begin{aligned} 5 - 7 \cdot 4 + 2(4 + 6) &= 5 - 7 \cdot 4 + 2(10) \\ &= 5 - 7 \cdot 4 + 20 \\ &= 5 - 28 + 20 \\ &= -23 + 20 \\ &= -3 \end{aligned}$$

To simplify an expression, you may need to simply exponents. To multiply exponents with the same base, add the exponents. To divide exponents with the same base, subtract the exponents.

$$y^a \times y^b = y^{a+b}$$

$$y^a \div y^b = y^{a-b}$$

SKILLS TIP

When simplifying an algebraic expression using the order of operations, multiplication and division must be performed from left to right. It's a mistake to assume that you must multiply before you divide. This is also true for the operations of addition and subtraction.

Evaluating Expressions

SKILLS TIP

The skills used for evaluating an expression can also be used when you are given formulas for various real-world phenomena. Simply plug in the given values, and then simplify in order to find the missing part(s).

Evaluate Expressions

Evaluating an algebraic expression involves substituting or replacing each variable with a number and then following the order of operations to simplify as much as possible.

Evaluate the expression $4xy - 5y$ if $x = 3$ and $y = 2$.

ANSWER: $4xy - 5y = 4 \cdot 3 \cdot 2 - 5 \cdot 2$ Substitute 3 for x and 2 for y.
$= 24 - 10$ Multiply.
$= 14$ Subtract.

If the numbers that are substituted for the variables are negative, you must use parentheses around the number being substituted. This ensures that the negative is used correctly when the expression is simplified.

Evaluate the expression $6xy + 2x - 3y$ if $x = -2$ and $y = -1$.

ANSWER: $6xy + 2x - 3y = 6(-2)(-1) + 2(-2) - 3(-1)$ Substitute -2 for x and -1 for y.
$= 12 + (-4) - (-3)$ Multiply.
$= 12 - 4 + 3$ Simplify.
$= 8 + 3$ Subtract.
$= 11$ Add.

Scientific notation uses the powers of ten to write very large or very small numbers. To write a number in scientific notation, use a number and a power of ten. The exponent indicates how many zeros are in the number or how many places come after the number in the ones place.

When writing a very large number in scientific notation, count the number of places you have to move a decimal point until there is one digit to the left of the decimal. The number of places you moved the decimal is the exponent.

When the exponent is positive, move the decimal point to the right and add zeros. The number becomes larger. When the exponent is negative, the number is smaller. Move the decimal point to the left and add zeros after the decimal point.

$3.9 \times 10^5 = 390{,}000$

$7.86 \times 10^{-3} = 0.00786$

Complete the activities below to check your understanding of the lesson content.

1. Evaluate the expression $5xy - z$ if $x = 4$, $y = -2$, and $z = -3$.

   A −43
   B −37
   C 15
   D 21
   E 43

Questions 2 and 3 refer to the rectangle below.

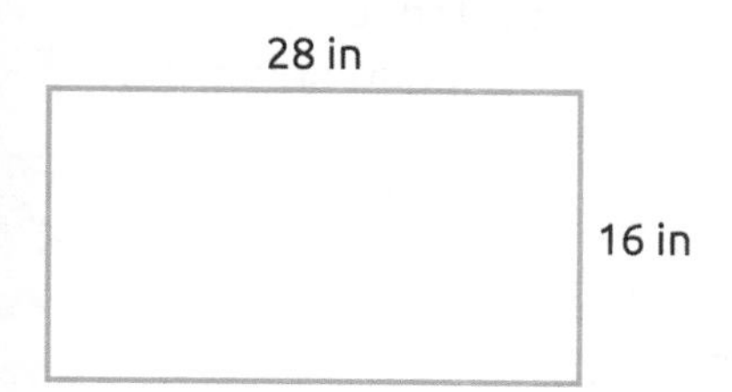

2. The formula for the perimeter of a rectangle is $P = 2l + 2w$. What is the perimeter of the rectangle shown here?

   A 1,792 inches
   B 448 inches
   C 224 inches
   D 88 inches
   E 44 inches

3. The formula for the area of a rectangle is $A = lw$. What is the area of the rectangle shown?

   A 1,792 in^2
   B 448 in^2
   C 224 in^2
   D 88 in^2
   E 44 in^2

4. To convert a temperature from Celsius to Fahrenheit, use the formula $F = 1.8C + 32$, where F is degrees Fahrenheit and C is degrees Celsius. What is the temperature in degrees Fahrenheit when the temperature is −5°C?

   A −41°F
   B 10.8°F
   C 23°F
   D 28.8°F
   E 41°F

5. Evaluate the expression $-3abc + c$ if $a = -1$, $b = 4$, and $c = -2$.

   A −26
   B −22
   C −11
   D 22
   E 26

6. Shama is driving on the highway and has her cruise control set for 61 miles per hour. She drives at this speed for 1.25 hours. Using the formula $d = rt$, how many miles did she travel?

   A 40.7 miles
   B 48.8 miles
   C 59.75 miles
   D 62.25 miles
   E 76.25 miles

KEY POINT!

Use parentheses when substituting negative numbers. Carefully follow the order of operations when simplifying.

TEST STRATEGY

First, try to solve the problem without looking at the answer choices. Often, the answer choices take into account common errors you might make when solving the problem.

Lesson Practice

KEY POINT!

Identify the given value for each variable in a formula, substitute each value into the formula, and solve.

7. The formula for the area of a trapezoid is $A = \frac{1}{2}h(b_1 + b_2)$, where h is the height of the trapezoid, b_1 is the length of the first base of the trapezoid, and b_2 is the length of the other base. What is the area of the trapezoid shown here?

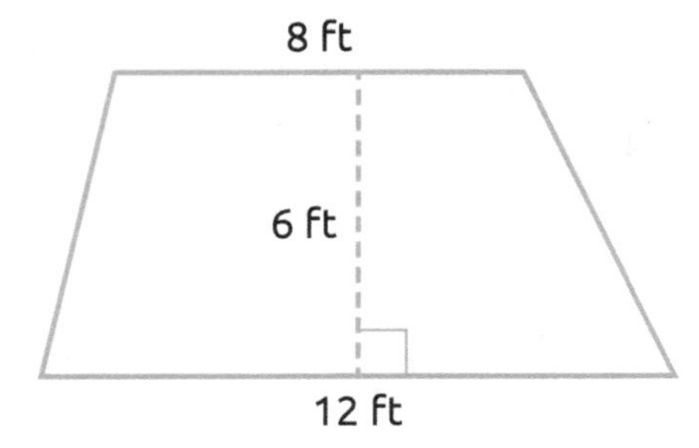

   A 23 ft²
   B 26 ft²
   C 60 ft²
   D 120 ft²
   E 144 ft²

8. The area of a circle is found using the formula $A = \pi r^2$, where r is the radius of the circle. What is the area of a circle that has a radius of 5 centimeters? Use $\pi = 3.14$.

   A 1.6 cm²
   B 15.7 cm²
   C 28.14 cm²
   D 78.5 cm²
   E 246.49 cm²

9. Earth orbits the sun at a rate of about 67,000 miles per hour. What is the speed in scientific notation?

   A 6.7×10^3
   B 6.7×10^4
   C 6.7×10^5
   D 67×10^4
   E 67×10^{10}

See page 90 for answers and help.

Performing Operations on Polynomials and Rational Expressions

Polynomials

A **monomial** is a single term consisting of a constant multiplied by a variable raised to a whole number power. A **polynomial** consists of several different monomials that are added or subtracted together. A **coefficient** of a term is the number that is being multiplied by the variable. In $5x^3$, the coefficient is 5.

Like Terms

Polynomials can be added or subtracted by combining like terms. **Like terms** are terms that look the same, such as terms that have the same variables and exponents. To add or subtract polynomials, combine any like terms by adding or subtracting only the coefficients of the like terms. The variables and the powers of the exponents do not change.

Add like terms: $3x^2y + 2xy - 5x + 4xy + y - 4y + 4x^2y$

ANSWER: One way to combine like terms is to group like terms to make it simpler to combine.

$$\begin{array}{r} 3x^2y + 2xy - 5x + 1y \\ +4x^2y + 4xy \qquad - 4y \\ \hline 7x^2y + 6xy - 5x - 3y \end{array}$$

Multiply Polynomials

To multiply polynomials, use the Distributive Property. When multiplying single terms that use the same variable, multiply the coefficients and add the exponents. Remember to combine any like terms.

Multiply: $(x + 3)(x - 4)$

ANSWER:

$$(x+3)(x-4)$$

$x^2 - 4x + 3x - 12$ FOIL

$x^2 - x - 12$ Combine like terms.

Multiply: $(x + 1)(x^2 + 2x - 5)$

ANSWER:

$$\begin{array}{r} x^3 + 2x^2 - 5x \\ x^2 + 2x - 5 \\ \hline x^3 + 3x^2 - 3x - 5 \end{array}$$

Distribute the x.

Distribute the 1 and line up like terms.

Combine like terms.

Performing Operations on Polynomials and Rational Expressions

SKILLS TIP

You can check your factorization by multiplying the factors together. If the product is the same as the original problem, you have factored it correctly.

Greatest Common Factor

Factoring polynomials is rewriting a polynomial as the product of all of its factors. The first thing to factor is the **greatest common factor (GCF)**, if there is one. The GCF is the greatest factor that is common to all of the terms in a polynomial.

Factor the GCF: $4x^3y + 2xy^2 - 6xy + 4xy^3 + 12x^2y^2$.

ANSWER: The GCF is $2xy$.

$$2xy\left(\frac{4x^3y}{2xy} + \frac{2xy^2}{2xy} - \frac{6xy}{2xy} + \frac{4xy^3}{2xy} + \frac{12x^2y^2}{2xy}\right)$$

Factor out the GCF by dividing each term by the GCF:

Then, simplify each term.

$2xy(2x^2 + y - 3 + 2y^2 + 6xy)$

SKILLS TIP

To check the factorization, FOIL and then distribute the GCF.

Factor Polynomials

To factor a polynomial in the form $ax^2 + bx + c$ where $a = 1$, the following steps can be used.

Three steps to follow:	Factor: $x^2 - 3x - 18$
1) Write two sets of parentheses. Put an x in each.	**1)** $(x \quad)(x \quad)$
2) Determine the signs: If: / Then: + + / + + − + / − − + − / + − − − / + −	**2)** Since the signs are − −, the signs in the factors will be + −. $(x + \quad)(x - \quad)$
3) Find the two numbers. Look for factors of c that add up to b.	**3)** The factors of −18 that add up to −3 are −6 and +3. $(x + 3)(x - 6)$

SKILLS TIP

In any problem where factoring is involved, always check to see if there is a GCF that can be factored out first. Sometimes this can make the difference in whether a polynomial can be factored.

Sometimes a GCF can be factored out of the polynomial before completely factoring the expression.

Factor: $2x^3 - 6x^2 + 4x$

ANSWER:

$2x^3 - 6x^2 + 4x$

$2x(x^2 - 3x + 2)$ Factor out the GCF.

$2x(x \quad)(x \quad)$ Write parentheses with xs.

$2x(x - \quad)(x - \quad)$ Determine the signs. $-+$ means $--$.

$2x(x - 1)(x - 2)$ Factors of 2 that add to -3 are -1 and -2.

Performing Operations on Polynomials and Rational Expressions

Rational Expressions

A **rational expression** is a fraction in which the numerator and/or the denominator are polynomials. Often, the polynomials in a rational expression can be factored and simplified. If the same factor appears in both the numerator and denominator, the factors reduce to 1 or cancel each other out.

Simplify: $\frac{x^2+3x-4}{x^2-5x+4}$

ANSWER: Begin by factoring the numerator and denominator. Then, cancel any identical factors that appear in the numerator and denominator.

$$\frac{x^2+3x-4}{x^2-5x+4}=\frac{(x+4)\cancel{(x-1)}}{(x-4)\cancel{(x-1)}}$$
$$=\frac{(x+4)}{(x-4)}$$

Some rational expressions have a GCF that can be factored out and canceled while simplifying.

Simplify: $\frac{2x^3+4x^2-30x}{2x^3-20x^2+42x}$

ANSWER: Begin by factoring out the GCF in the numerator and denominator.

$$\frac{2x^3+4x^2-30x}{2x^3-20x^2+42x}=\frac{2x(x^2+2x-15)}{2x(x^2-10x+21)}$$
$$=\frac{\cancel{2x}(x+5)\cancel{(x-3)}}{\cancel{2x}(x-7)\cancel{(x-3)}}$$
$$=\frac{(x+5)}{(x-7)}$$

Lesson Practice

Complete the activities below to check your understanding of the lesson content.

1. What is the simplified form of the polynomial shown here?

   $7xy^2+4xy-2x-6xy+3xy^2+5x$

   A $11x^6y^6$

   B $10xy^2+10xy+7x$

   C $10xy^2+10xy-3x$

   D $10xy^2-2xy-3x$

   E $10xy^2-2xy+3x$

2. Which of the following is the difference of the polynomials shown here?

   $(5x^2+2x-6)-(7x^2+3x+4)$

   A $12x^2+5x-2$

   B $-2x^2-x-10$

   C $-2x^2+5x-2$

   D $-2x^2-x-2$

   E $-2x^2+5x-10$

KEY POINT!

When you have a subtraction sign (negative sign) directly before a quantity in parentheses, the negative must be distributed to each term in the parentheses before like terms are combined.

Lesson Practice

Questions 3 and 4 refer to the information below.

$(x+3)$ in

$(2x-5)$ in

3. The perimeter of a rectangle can be found using the formula $P = 2l + 2w$. Which of the following represents the perimeter of the rectangle shown here?
   - A $6x - 4$ inches
   - B $3x - 2$ inches
   - C $6x - 14$ inches
   - D $2x^2 + x - 15$ inches
   - E $2x^2 + 11x - 15$ inches

4. The area of a rectangle can be found using the formula $A = lw$. Which of the following represents the area of the rectangle shown?
   - A $6x - 4$ in^2
   - B $3x - 2$ in^2
   - C $6x - 14$ in^2
   - D $2x^2 + x - 15$ in^2

5. Simplify: $\frac{x^2 - 3x - 4}{x^2 - 5x + 4}$
   - A $\frac{1}{2x}$
   - B $\frac{x-1}{x+1}$
   - C $\frac{x+1}{x-1}$
   - D $\frac{x-4}{x+4}$
   - E $\frac{x+4}{x-4}$

See page 90 for answers and help.

6. Factor the following expression: $3x^3 - 3x^2 - 18x$.
   - A $3x(x^2 - x - 6)$
   - B $3x(x + 1)(x - 6)$
   - C $3x(x + 6)(x - 1)$
   - D $3x(x + 2)(x - 3)$
   - E $3x(x - 2)(x + 3)$

7. The student body consists of $x + 17$ boys and $3x - 34$ girls. If $x - 15$ students leave after the first semester, write an expression for the number of students who will remain in the student body for the second semester.
   - A $-3x - 32$
   - B $3x - 2$
   - C $3x - 32$
   - D $4x - 17$
   - E $5x - 2$

8. The area of a square can be found using the formula $A = s^2$, where s is the length of a side of the square. What is the area of the square shown below?

   $(x+3)$ ft

   - A $x^2 + 9$ ft^2
   - B $x^2 + 3x + 9$ ft^2
   - C $x^2 + 6x + 9$ ft^2
   - D $4x + 12$ ft^2
   - E $2x + 6$ ft^2

TEST STRATEGY

When time allows, go back and check your answers. Multiply the factors in your answer together to see if you get the original polynomial.

Solving Multi-Step Equations

Equations usually have one variable with an equal sign and sometimes take many steps to solve. Generally, each equation can be solved following these steps in order.

1) Simplify the parentheses on both sides of the equal sign. This often involves using the Distributive Property.
2) Combine like terms on each side of the equal sign.
3) Move the terms with variables to one side of the equal sign and the constants to the other side of the equal sign.
4) Isolate the variable.

Solve: $11x - (2 - 4x) = 13 + 2(x - 1)$

ANSWER: $11x - (2 - 4x) = 13 + 2(x - 1)$

$11x - 2 + 4x = 13 + 2x - 2$	Distribute -1 and 2.
$15x - 2 = 2x + 11$	Combine like terms.
$13x - 2 = 11$	Get terms with variables on one side.
$13x = 13$	Get constants on other side.
$x = 1$	Isolate the variable.

SKILLS TIP

When the steps for solving a multi-step equation are not followed in order, the process can introduce fractions unnecessarily and become complicated.

Solving Quadratic Equations by Factoring

A **quadratic equation** is an equation whose greatest exponent is 2. It can usually be solved by factoring and using the Zero-Product Property. The **Zero-Product Property** states that if the product of two factors is 0, then one or both factors must be equal to 0. So, to solve a quadratic equation, factor and then set each factor equal to 0. Then, isolate the variable for each factor.

Solve: $x^2 - 5x - 36 = 0$

ANSWER: $x^2 - 5x - 36 = 0$

$(x - 9)(x + 4) = 0$		Factor.
$x - 9 = 0$	$x + 4 = 0$	Set both factors equal to 0.
$x = 9$	$x = -4$	Solve for x.

Solving Equations and Inequalities

Using the Quadratic Formula to Solve Equations

Quadratic equations written in the form $ax^2 + bx + c$ can also be solved using the quadratic formula, which states $x = \frac{-b \pm \sqrt{b^2 - 4ac}}{2a}$. Though the formula seems complex, it can be solved by substituting and simplifying.

Solve: $2x^2 - 13x + 20$

ANSWER: $2x^2 + (-13)x + 20$ Rewrite in $ax^2 + bx + c$ form.

$x = \frac{-b \pm \sqrt{b^2 - 4ac}}{2a}$ Use Quadratic Formula.

$x = \frac{-(-13) \pm \sqrt{(-13)^2 - 4(2)(20)}}{2(2)}$ Substitute: $a = 2$, $b = -13$, $c = 20$.

$x = \frac{13 \pm \sqrt{169 - 160}}{4}$ Simplify.

$x = \frac{13 + 3}{4}, x = \frac{13 - 3}{4}$ Identify both solutions.

$x = 4, x = \frac{5}{2}$ Simplify.

Solve Inequalities

An **inequality** is similar to an equation except that it has a <, >, ≤, or ≥ instead of an equal sign. An inequality usually has an infinite number of solutions. A number line graph can be used to show the solutions to an inequality. It is important to understand the difference between < or > and ≤ or ≥.

Symbols	What They Mean	Symbol on Number Line
< or >	Less than or greater than a number, but the number is NOT included. So, $x > 1$ means that x is any number greater than 1 and does not include 1.	Use an open circle to plot the number because the number is not included in the solution.
≤ or ≥	Less than or equal to or greater than or equal to a number, and the number IS included. So, $x \geq 1$ means any number equal to or greater than 1.	Use a closed circle to plot the number because the number is included in the solution.

Solve: $2x + 9 + 3x + 1 > 0$

ANSWER:

$2x + 9 + 3x + 1 > 0$

$5x + 10 > 0$ Combine like terms

$5x > -10$ Subtract 10 from both sides.

$x > -2$ Divide both sides by 5.

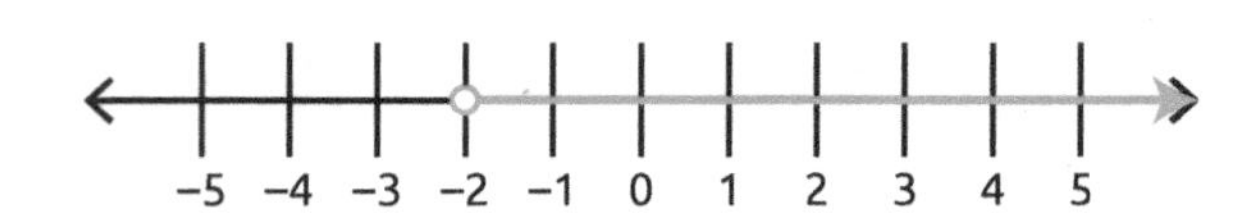

When solving an inequality, if you multiply or divide both sides of the inequality

SKILLS TIP

The steps for solving an inequality are identical to those for solving an equation. Work to isolate the variable on one side of the inequality symbol. Remember to flip the symbol when multiplying or dividing both sides by a negative number.

by a negative number, then you must flip the inequality symbol.

Solve: $5x - 3 - 6x + 1 \geq 0$

ANSWER: $5x - 3 - 6x + 1 \geq 0$

$-x - 2 \geq 0$ Combine like terms

$-x \geq 2$ Add 2 to both sides.

$x \leq -2$ Divide both sides by -1 and flip $\geq$.

Solving Cubic Equations

A cubic equation is an equation whose greatest exponent is 3. Some solutions to cubic equations may involve a special value i, which equals $\sqrt{-1}$. Since $i = \sqrt{-1}$, $i^2 = -1$, and substituting i^2 for -1 can be useful when finding solutions to cubic equations. While it can be challenging to solve a cubic equation directly, easier methods exist for checking the solutions to a cubic equation.

Which cubic equation has 3 and $4i$ as solutions:
$x^3 - 3x^2 - 16x + 48$ or $x^3 - 3x^2 + 16x - 48$?

ANSWER: If a cubic equation has 3 as a solution, then it has $(x - 3)$ as a factor. First, determine which of the cubic equations has $(x - 3)$ as a factor. Since $x^3 - 3x^2 - 16x + 48 = (x - 3)(x^2 - 16)$, and $x^3 - 3x^2 + 16x - 48 = (x - 3)(x^2 + 16)$, both equations have 3 as a solution. To determine which equation has $4i$ as a solution, continue factoring both equations.

$x^3 - 3x^2 - 16x + 48 = (x - 3)(x^2 - 16)$

$= (x - 3)(x + 4)(x - 4)$ Difference of squares: $a^2 - b^2 = (a + b)(a - b)$

Since the equation $x^3 - 3x^2 - 16x + 48$ has the factors $(x - 3)$, $(x + 4)$, and $(x - 4)$, it has the solutions 3, –4, and 4. This equation does not have 3 and $4i$ as solutions.

$x^3 - 3x^2 + 16x - 48 = (x - 3)(x^2 + 16)$

$= (x - 3)(x^2 - (-16))$ Rewrite equation.

$= (x - 3)(x^2 - (16i^2))$ Substitute: $-1 = i^2$

$= (x - 3)(x + 4i)(x - 4i)$ Difference of squares: $a^2 - b^2 = (a + b)(a - b)$

Since the equation $x^3 - 3x^2 + 16x - 48$ has the factors $(x - 3)$, $(x + 4i)$, and $(x - 4i)$, it has the solutions 3, $-4i$, and $4i$. This equation does have 3 and $4i$ as solutions.

Lesson Practice

Complete the activities below to check your understanding of the lesson content.

1. Consider the equation $3x + 4 = 2(x - 4)$. What is the first step that should be taken to solve the equation?

   A Subtract 4 from both sides of the equation.

   B Divide both sides of the equation by 3.

   C Distribute the 2 into the parentheses.

   D Add 4 to both sides of the equation.

   E Subtract 2 from both sides of the equation.

2. Consider the equation $5x - 2 = 4x + 3$. Which of the following steps would be done when solving this equation?

   A. Divide both sides of the equation by 5.
   B. Combine the $5x$ and $4x$ terms together.
   C. Combine the -2 and $+3$ terms together.

   A A only

   B B only

   C C only

   D A and B only

   E B and C only

3. Which cubic equation has -2 and $3i$ as solutions?

   A $x^3+2x^2+9x+18=0$

   B $x^3+2x^2-9x-18=0$

   C $x^3+2x^2+9x-18=0$

   D $x^3-2x^2-9x+18=0$

   E $x^3-2x^2-9x-18=0$

Questions 4 and 5 refer to the information below.

The perimeter of a triangle is 250 centimeters. The sides of the triangle are represented by x, $x-2$, and $x+12$.

4. Which equation can be used to find the length of each side of the triangle?

   A $(x)(x-2)(x+12)=250$

   B $(x-2)(x+12)=250$

   C $x+(x+10)=250$

   D $x+(x-2)+(x+12)=250$

   E $(x-2)+(x+12)=250$

5. What is the length of each side of the triangle?

   A 1 cm, 10 cm, and 25 cm

   B 78 cm, 80 cm, and 92 cm

   C 60 cm, 90 cm, and 100 cm

   D 70 cm, 80 cm, and 100 cm

   E 100 cm, 90 cm, and 25 cm

6. What are the real number solutions to the following quadratic equation?

   $x^2+3x-10=0$

   A -2 and 5

   B -5 and 2

   C 5 and 2

   D -1 and 10

   E -10 and 1

7. What is the solution to $3x+7-8x+3>0$?

   A

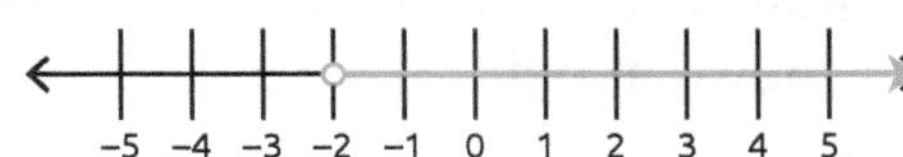

   B

   C

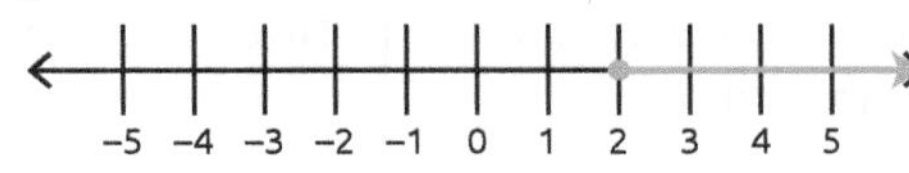

   D

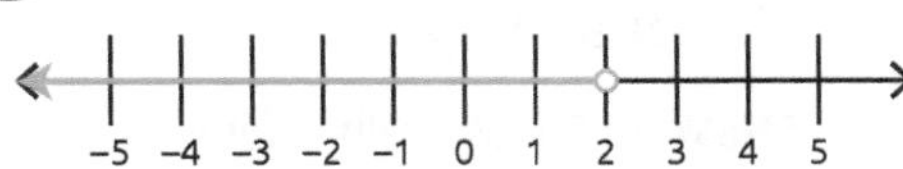

   E

8. Terrelle earned \$778.36 last week for working 53 hours. He gets paid his hourly rate for the first 40 hours, 1.5 times his regular hourly rate for the next 10 hours, and double his regular hourly rate for any time over that. His earnings last week can be represented by the equation $40x+10(1.5x)+3(2x)=\$778.36$, where x is his hourly rate. What is his hourly rate of pay?

   A \$10.96 per hour

   B \$11.79 per hour

   C \$12.76 per hour

   D \$13.08 per hour

   E \$14.69 per hour

See page 90 for answers and help.

Graphing Equations and Inequalities

Linear Graphing

A linear equation will form a line on the **Cartesian coordinate plane**, which is shown at right.

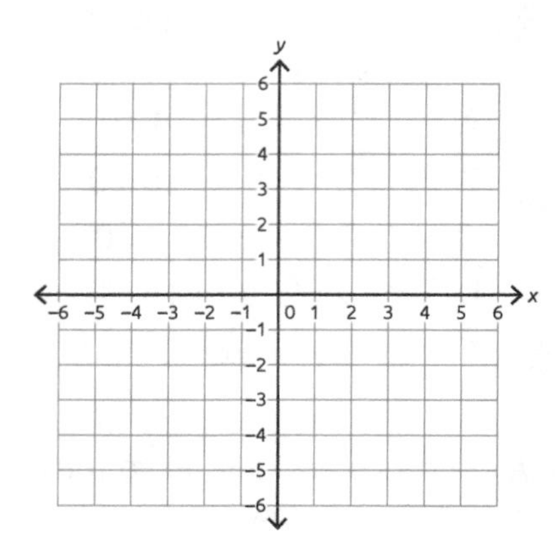

A linear equation can be graphed in two ways. The first is to create a table of values. Choose any three values of *x* that are easy to work with when substituted into the equation to find each corresponding *y*-value. Plot the three points on the plane, and then connect the points to form the line.

Use a table of values to graph the line formed by this equation: $y = 2x - 1$

ANSWER: Choose values for *x* like −1, 0, or 1. Find the corresponding values of *y*, and then plot and connect the points.

x	*y*
−1	−3
0	−1
1	1

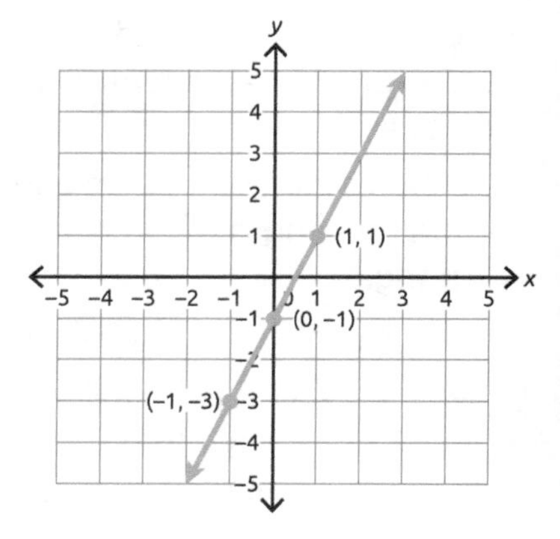

Each line has points that are called **intercepts**, points where the line intersects the axes. The ***x*-intercept** is where the line crosses the *x*-axis, which occurs when $y = 0$. The ***y*-intercept** is where the line crosses the *y*-axis, which occurs when $x = 0$. A line can be graphed using the intercepts. Simply find *y* when $x = 0$, and then find *x* when $y = 0$. These two points can be used to graph the line.

Use a table with the intercepts to graph the line formed by the equation $y = \frac{1}{2}x + 3$.

ANSWER: Use $x = 0$ and then $y = 0$ in the table.

x	*y*
−1	3
−6	0

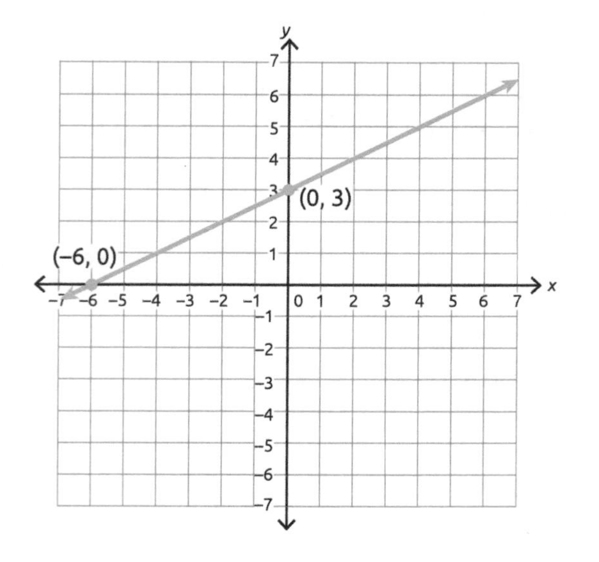

Slope-Intercept Form

The **slope-intercept form** of the equation of a line is an equation in the form $y = mx + b$, where m is the slope, and b is the y-intercept of the line. The **slope** of a line measures the steepness of the line. It can be thought of as the $\frac{\text{rise}}{\text{run}}$ of the line. If the slope is positive, the line goes up from left to right. If the slope is negative, the line goes down from left to right.

To graph the equation of a line in this form, start by plotting the y-intercept, and then use the rise and run of the slope to graph one or two more points.

Use the slope and y-intercept to graph the line formed by this equation:
$y = -\frac{3}{4}x + 4$

ANSWER: Plot the y-intercept at (0, 4). Then, use the rise of −3 and the run of +4 or the rise of +3 and a run of −4 to plot two other points on the line.

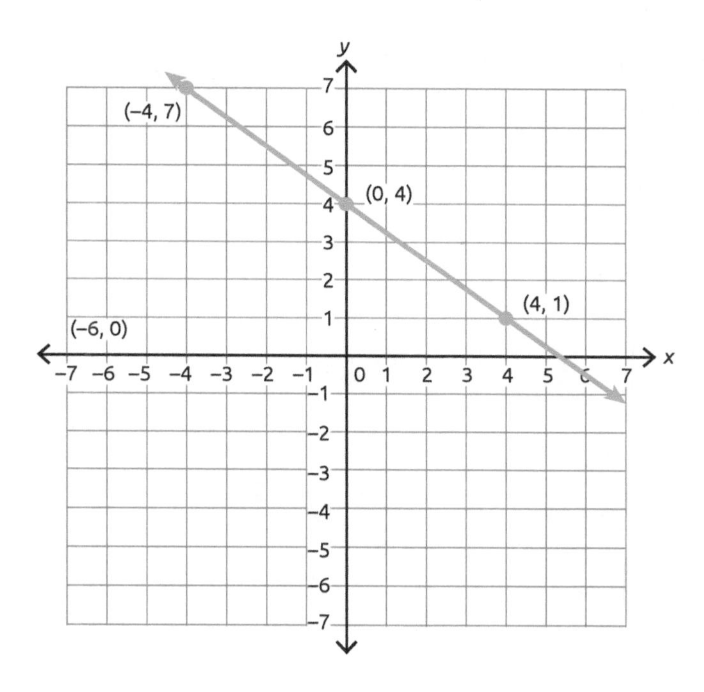

SKILLS TIP

When the slope is negative, either the rise can be negative and the run can be positive, or the rise can be positive and the run can be negative. If the slope is positive, either both rise and run are positive or both rise and run are negative.

Graphing Linear Inequalities

A linear inequality can be graphed in a way that is similar to how a linear equation is graphed. Use the y-intercept and the slope to determine the location of the points that are on the line.

If the inequality symbol is $<$ or $>$, then use a dashed line because the line is not included in the solution. If the inequality symbol is $\leq$ or $\geq$, then use a solid line because the line is included in the solution. Then, shade the area either above or below the line. If the inequality begins with $y <$ or $y \leq$, shade below the line. If the inequality begins with $y >$ or $y \geq$, shade above the line.

Graphing Equations and Inequalities

Graph the solution: $y < \frac{2}{3}x + 1$

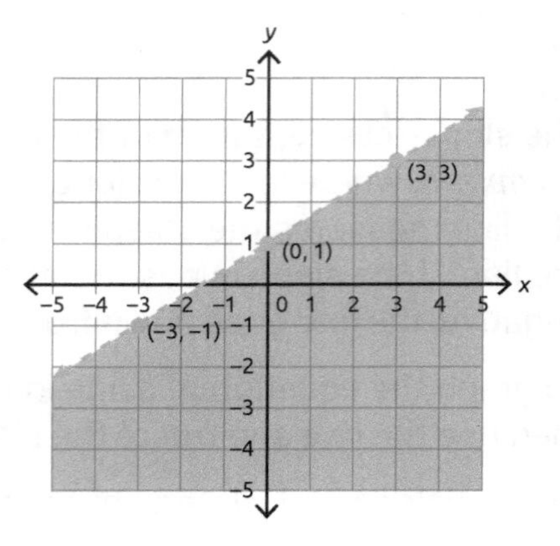

ANSWER: Graph the line $y = \frac{2}{3}x + 1$ as a dashed line and shade the area below the line.

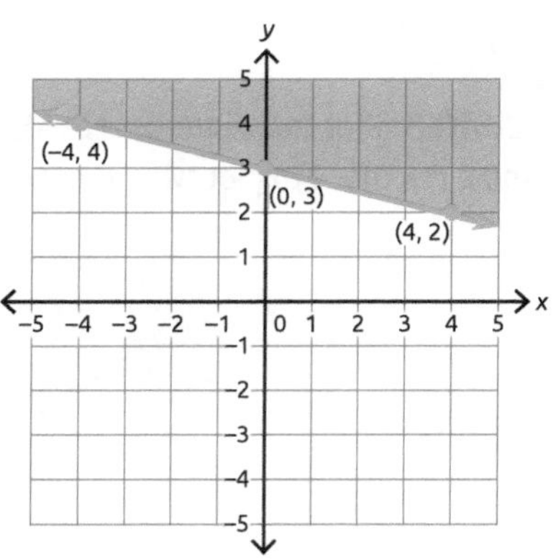

Graph the solution: $y \geq -\frac{1}{4}x + 3$

ANSWER: Graph the line $y = -\frac{1}{4}x + 3$ as a solid line and shade above the line.

Complete the activities below to check your understanding of the lesson content.

Questions 1–3 refer to the information below.

The diagram shows a graph of the line $y = -2x + 4$.

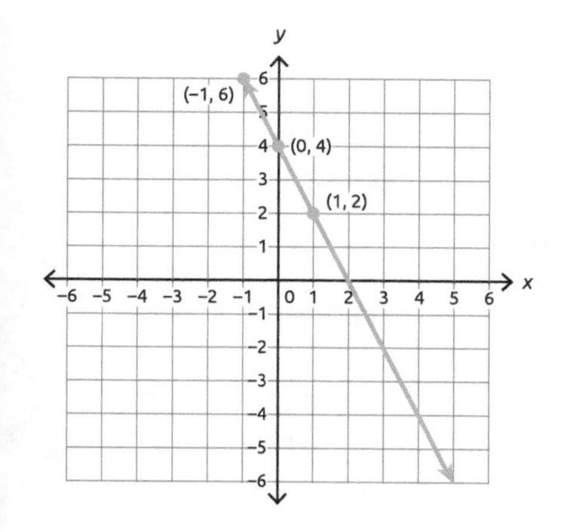

1. What is the slope of the line?

   A $\frac{1}{4}$

   B 4

   C $-\frac{1}{2}$

   D −2

   E −4

2. At what point does the line cross the *x*-axis?

   A (−4, 0)

   B (−2, 0)

   C (2, 0)

   D (4, 0)

   E (12, 0)

3. At what point does the line cross the *y*-axis?

   A (0, −4)

   B (0, −2)

   C (0, 2)

   D (0, 4)

   E (0, 12)

KEY POINT!

The slope-intercept form of the equation of a line is $y = mx + b$, where the coefficient of the x term is the slope of the line. The constant, b, is the y-intercept.

TEST STRATEGY

Look at the sign of the slope to determine the direction of the line on the graph. Look at the *y*-intercept in the equation and the location of it on the graph. These things can be helpful in identifying the correct graph.

Lesson Practice

4. Which inequality is represented in the graph shown below?

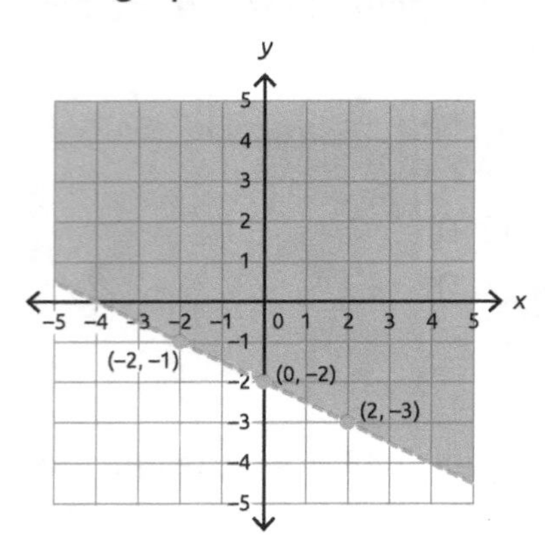

A $y \geq -\frac{1}{2}x - 2$

B $y > -\frac{1}{2}x - 2$

C $y \leq -\frac{1}{2}x - 2$

D $y < -\frac{1}{2}x - 2$

E $y > \frac{1}{2}x - 2$

5. Which of the following represents the graph of the line $y = -2x - 1$?

A

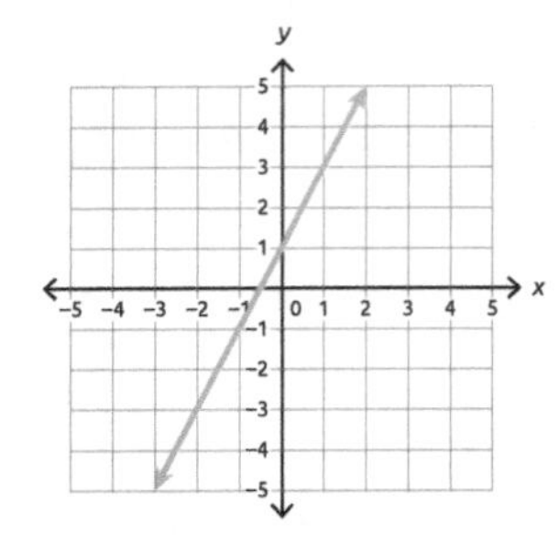

B

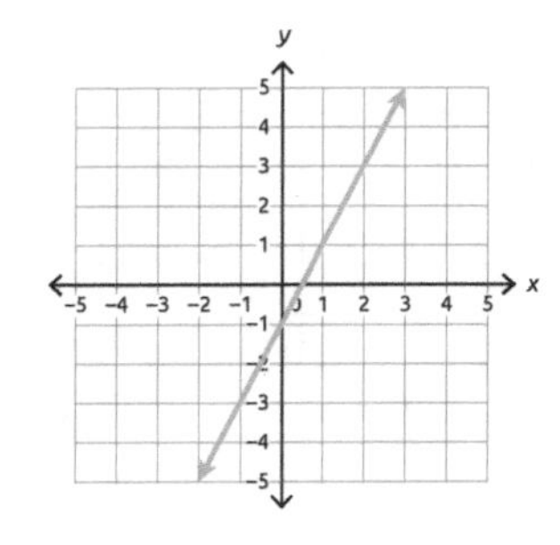

C

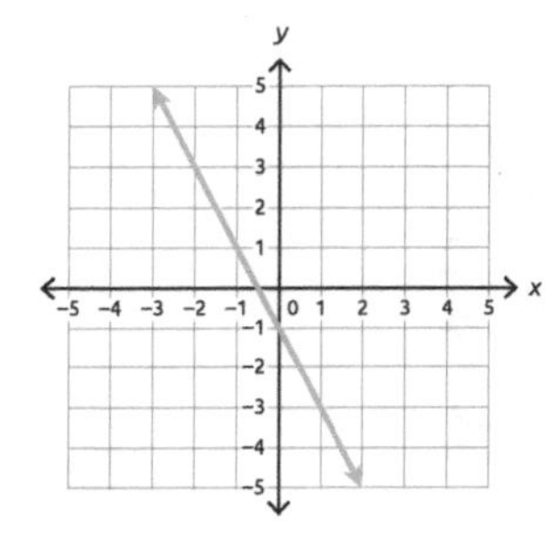

D

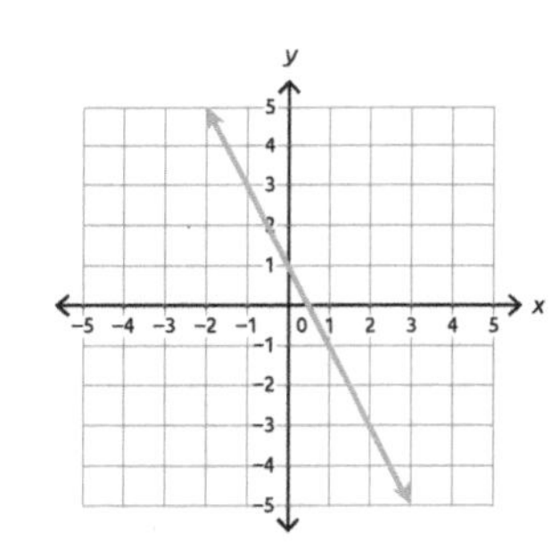

E

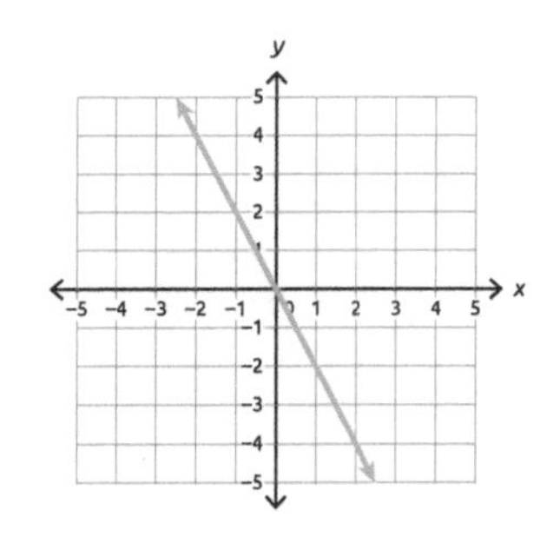

See page 90 for answers and help.

Systems of Equations

A **system of equations** is two or more linear equations in which the same variables work together simultaneously. To solve a system of linear equations, such as the one shown here, determine the ordered pair(s), (x, y), that make both equations true.

$$\begin{cases} 2x+3y=1 \\ x-2y=-3 \end{cases}$$

A linear system of equations can be solved in three ways: graphing, substitution, and elimination.

Solving Systems of Equations by Graphing

To solve a system of equations by graphing, graph both lines, and then find their point of intersection.

Solve the system of equations by graphing:

$$\begin{cases} 4x+y=-3 \\ -3x+y=4 \end{cases}$$

ANSWER: First, solve each equation for y. Then, use the slope and the y-intercept to graph each line, and then identify the point of intersection.

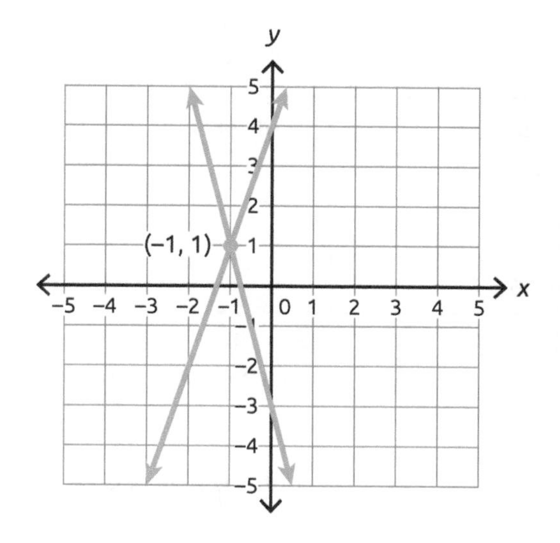

The point of intersection is at $(-1, 1)$. This is the only solution that makes both equations true.

A system of linear equations has three possibilities for the number of solutions. Most systems, when graphed, intersect at exactly one point and, therefore, have exactly one solution. However, the two lines could be parallel and have the same slope. Since they never intersect, the system has no solution. If the lines are parallel, then there is no ordered pair that will make both equations true simultaneously. The two lines could also coincide, or they could be the same line. In this case, an infinite number of solutions could make both equations true.

Solving Systems of Equations

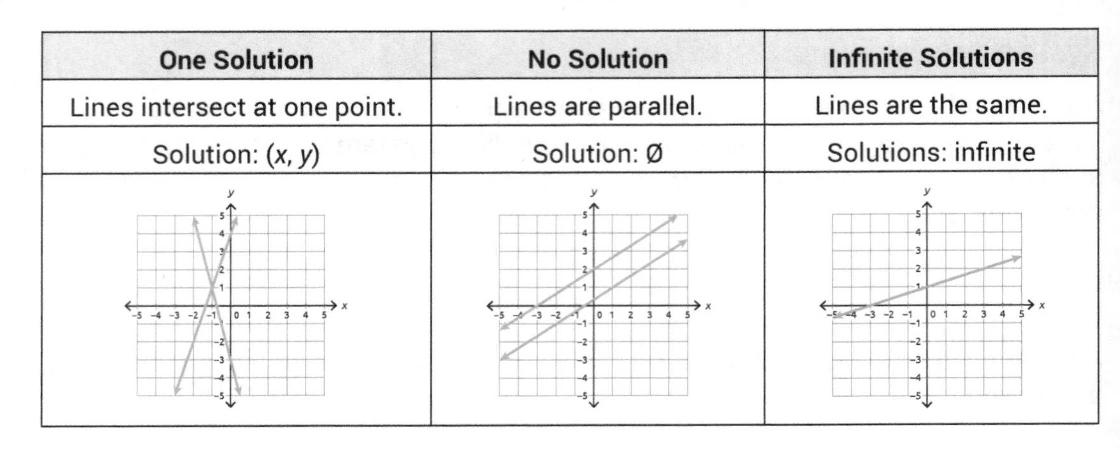

One Solution	No Solution	Infinite Solutions
Lines intersect at one point.	Lines are parallel.	Lines are the same.
Solution: (x, y)	Solution: Ø	Solutions: infinite

SKILLS TIP

Most linear systems of two equations have exactly one solution: the ordered pair where the lines intersect. Still, it is important to remember that occasionally they have no solution or have many (infinite) solutions.

Solving Systems of Equations by Substitution

If two quantities are equivalent, then one can be substituted for the other. Substitution can be used to solve systems of linear equations. Follow these four steps to solve a linear system by substitution.

Solve by substitution: $\begin{cases} -3x+2y=19 \\ 2x+y=-8 \end{cases}$	**Steps**
$y=-2x-8$	Solve one equation for either variable. The y in the bottom equation is the easiest to isolate because its coefficient is 1.
$-3x+2(-2x-8)=19$	Substitute $-2x - 8$ for y in the top equation.
$-3x+2(-2x-8)=19$ $-3x-4x-16=19$ $-7x-16=19$ $-7x=35$ $x=-5$	Solve for x.
$2(-5)+y=-8$ $-10+y=-8$ $y=2$	Substitute -5 for x in one of the original equations, and then solve for y.
$(-5, 2)$	Write the solution as an ordered pair.

Solving Systems of Equations by Elimination

Elimination is another method that can be used to solve systems of linear equations. To do this, modify one or both equations, if necessary, so that one of the variables is eliminated. Follow these four steps when solving a linear system by elimination.

Solve by elimination: $\begin{cases} 3x+4y=-6 \\ 2x+5y=-11 \end{cases}$	**Steps**
$\begin{cases} 3x+4y=-6 \\ 2x+5y=-11 \end{cases}$	Make sure both equations are in standard form, $Ax+By=C$. If not, rewrite them in this form.
$\begin{cases} 2(3x+4y=-6) \\ -3(2x+5y=-11) \end{cases} \rightarrow \begin{cases} 6x+8y=-12 \\ -6x-15y=33 \end{cases}$ $-7y=21$ $y=-3$	Decide which variable to eliminate. Multiply one or both equations so that the coefficients of that variable are the same number with opposite signs. Then, add the new equations to eliminate that variable. Solve for the other variable.
$3x+4(-3)=-6$ $3x-12=-6$ $3x=6$ $x=2$	Substitute −3 for *y* in one of the original equations, and then solve for *x*.
(2, −3)	Write the solution as an ordered pair.

Systems of Equations with No Solution or Infinite Solutions

When solving systems of linear equations by substitution or elimination, there will sometimes be no solution (parallel lines) or infinitely many solutions (same lines). If the variable is eliminated, look at the statement that is remaining. If it is a true statement, then there are infinitely many solutions. If it is a false statement, then there is no solution.

Solve the system of equations by substitution: $\begin{cases} x+3y=1 \\ -3x-9y=3 \end{cases}$

ANSWER:

$$\begin{aligned} x &= -3y+1 \\ -3(-3y+1)-9y &= 3 \\ 9y-3-9y &= 3 \\ -3 &\neq 3 \end{aligned}$$

Since the terms with variables cancel out and the statement is false, there is no solution to this system.

Solve the system of equations by elimination: $\begin{cases} -2x+y=4 \\ 4x-2y=-8 \end{cases}$

ANSWER: $\begin{cases} 2(-2x+y=4) \\ 4x-2y=-8 \end{cases} \rightarrow \begin{cases} -4x+2y=8 \\ 4x-2y=-8 \end{cases}$

$$0=0$$

Since the terms with variables cancel out and the statement is true, there are infinite solutions to this system.

Lesson Practice

Complete the activities below to check your understanding of the lesson content.

Questions 1 and 2 refer to the information below.

A drama production charged $6 for adults and $4 for children. The production sold a total of $990 in tickets. The total number of tickets sold was 185.

1. Which linear system of equations can be used to find the number of adult tickets, a, and child tickets, c, that were sold?

   A $\begin{cases} a+c=990 \\ a+c=185 \end{cases}$

   B $\begin{cases} 6a+4c=185 \\ a+c=990 \end{cases}$

   C $\begin{cases} 6a+4c=990 \\ a+c=185 \end{cases}$

   D $\begin{cases} 4a+6c=990 \\ a+c=185 \end{cases}$

   E $\begin{cases} 6a+4c=990 \\ 6a+4c=185 \end{cases}$

2. How many adults and children attended the play?

   A 60 adults and 125 children
   B 125 adults and 60 children
   C 100 adults and 85 children
   D 85 adults and 100 children
   E 110 adults and 75 children

3. Which of the following is a reasonable first step when using elimination to solve this system of linear equations?

   $\begin{cases} 3x-y=11 \\ -2x+2y=-6 \end{cases}$

   A Multiply the top equation by 2 and the bottom by 3 to eliminate y.
   B Multiply the top equation by 3 and the bottom by 2 to eliminate x.
   C Multiply the bottom equation by 2 to eliminate y.
   D Multiply the top equation by 2 to eliminate y.
   E Multiply both equations by 2 to eliminate y.

4. Which of the following is a reasonable first step when using substitution to solve this system of linear equations?

   $\begin{cases} 3x-y=11 \\ -2x+2y=-6 \end{cases}$

   A Get y by itself in the top equation.
   B Substitute the first equation into the sec ond equation.
   C Substitute −6 into the top equation.
   D Multiply the top equation by 2.
   E Substitute 11 into the bottom equation.

KEY POINT!

To solve a system of linear equations, use substitution or elimination. Either one can be used to determine the solution.

5. Which of the following graphs represents the solution to this system of linear equations?

$$\begin{cases} y = 3x - 2 \\ y = x + 2 \end{cases}$$

A

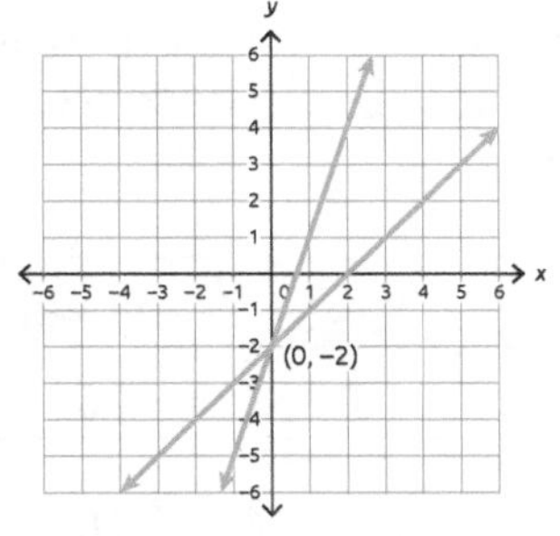

B

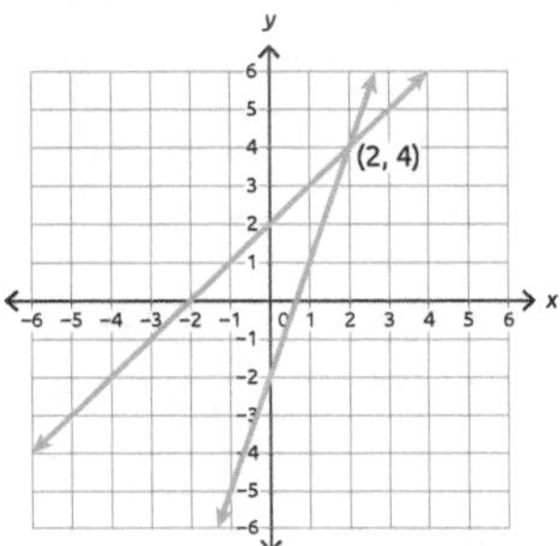

C

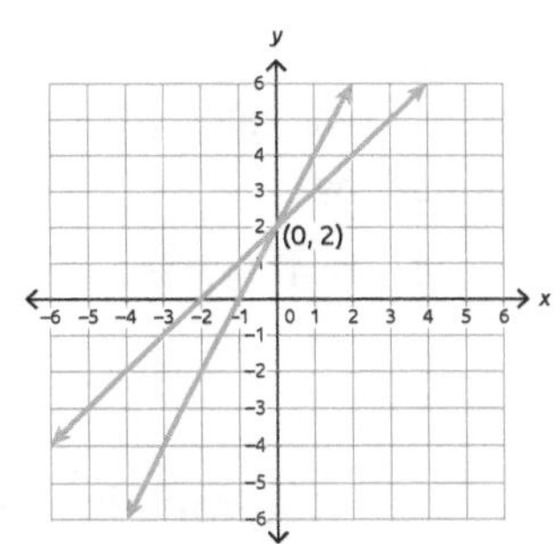

D

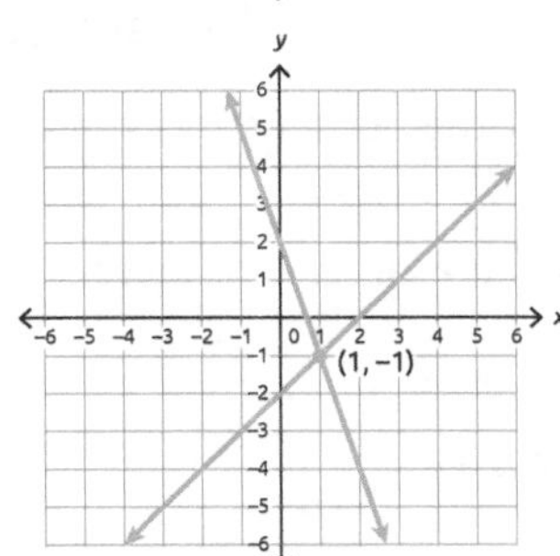

E

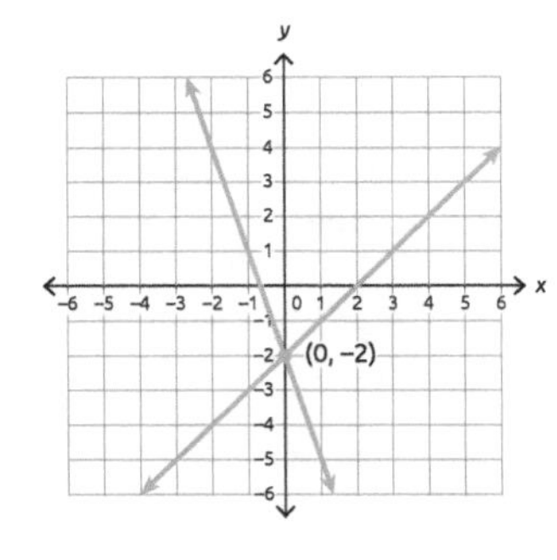

Questions 6 and 7 refer to the information below.

The sum of two numbers is 37, and their difference is 13.

6. Which linear system of equations can be used to find the two numbers?

A $\begin{cases} x - y = 37 \\ x + y = 13 \end{cases}$

B $\begin{cases} x + y = 37 \\ xy = 13 \end{cases}$

C $\begin{cases} xy = 37 \\ x - y = 13 \end{cases}$

D $\begin{cases} x + y = 37 \\ xy = 13 \end{cases}$

E $\begin{cases} x + y = 37 \\ x - y = 13 \end{cases}$

7. What are the two numbers?

A 20 and 17
B 32 and 5
C 18 and 19
D 25 and 12
E 22 and 15

8. What is the solution of this linear system of equations?

$$\begin{cases} x - 5y = -8 \\ -2x + 10y = 16 \end{cases}$$

A (2, 2)
B (−2, 2)
C (−3, 1)
D There is no solution.
E There are infinite solutions.

See page 90 for answers and help.

TEST STRATEGY

When examining graphs in the answer choices, pay close attention to the important parts of the graph such as the slope and the *y*-intercept.

KEY POINT!

If the terms with variables cancel and the remaining statement is true, then there are infinite solutions. If the terms with variables cancel and the remaining statement is false, then there is no solution.

Solving Problems Using Expressions, Equations, and Inequalities

Consecutive Integer Problems

Many types of word problems can help you understand how to use and solve equations. Consecutive integers are simple integers that directly follow each other, such as 1, 2, and 3 or −5, −4, and −3. Consecutive integer problems are worded so that you can use expressions to form equations that can be solved in order to determine the consecutive integers being described. These expressions are formed by patterns shown in the following table.

Key Words	Expressions	Examples
Consecutive Integers	$n, n+1, n+2$	0, 1, 2 or −10, −9, −8
Consecutive Even Integers	$n, n+2, n+4$ (n must be even)	0, 2, 4 or −10, −8, −6
Consecutive Odd Integers	$n, n+2, n+4$ (n must be odd)	1, 3, 5 or −7, −5, −3

The sum of three consecutive odd integers is 147. What are the three integers?

ANSWER: Since the three integers are consecutive odd integers, use n, $n+2$, and $n+4$, where n is an odd integer.

$n+(n+2)+(n+4)=147$ Write the equation.

$3n+6=147$ Combine like terms.

$3n=141$ Subtract 6 from each side.

$n=47$ Divide both sides by 3.

So, the three consecutive odd integers whose sum is 147 are 47, 49, and 51.

SKILLS TIP

When setting up a number sentence, it is important to be able to identify the key words in the word problem that indicate a mathematical operation. These key words help with the "translation" to form a number sentence.

Direct Translation Problems

Direct translation problems are problems about everyday situations involving numbers that must be translated into number sentences and then solved.

A patio has a length that is twice as long as its width. If the perimeter of the patio is 48 feet, then what are the dimensions of the patio?

ANSWER: Use the formula $P = 2l + 2w$. Let $l = 2w$, then substitute $2w$ for l and solve.

$P=2l+2w$ Use the formula.

$48=2(2w)+2w$ Substitute 48 for P and $2w$ for l.

$48 = 4w + 2w$ Multiply.

$48=6w$ Combine like terms.

$8=w$ Divide both sides by 6.

The width is 8 feet, and the length is twice the width, or 16 feet.

Four pens and three notebooks cost \$10.65. Three pens and four notebooks cost \$11.05. How much do five pens and five notebooks cost?

ANSWER: This problem has two unknowns—pens and notebooks—so two variables are needed. It has two totals, so two linear equations are needed. Use elimination to solve the linear system of equations.

$$\begin{cases} 4p + 3n = 10.65 \\ 3p + 4n = 11.05 \end{cases}$$ Set up both equations.

$$\begin{cases} 12p + 9n = 31.95 \\ -12p - 16n = -44.20 \end{cases}$$ Multiply the top by 3 and the bottom by −4.

$-7n = -12.25$ Add the equations to eliminate a variable.

$n = 1.75$ Divide both sides by −7.

So, the notebooks cost \$1.75. Substitute the value of n to find the cost of the pens.

$4p + 3(1.75) = 10.65$	Substitute 1.75 for n.
$4p + 5.25 = 10.65$	Multiply.
$4p = 5.40$	Subtract 5.25 from each side.
$p = 1.35$	Divide each side by 4.

So, the pens cost \$1.35. Use both values to determine how much 5 of each cost.

$5(1.35) + 5(1.75) = 15.50$

So, 5 pens and 5 notebooks cost \$15.50.

Solving Problems Using Inequalities

To solve problems involving inequalities, set up the inequalities as you would equations, but use inequality symbols instead of an equal sign.

Jackson is renting a car for \$140 per week plus \$0.20 per mile. How many miles can Jackson drive if he wants to spend a maximum of \$250 for one week?

ANSWER: Set up the inequality and solve.

$0.2m + 140 \leq 250$	Set up the inequality.
$0.2m \leq 110$	Subtract 140 from each side.
$m \leq 550$	Divide each side by 0.2.

Jackson can drive up to 550 miles during the week to spend at most \$250.

SKILLS TIP

Each unknown in a real-world situation is represented by a variable. The number of variables indicates the number of equations you will need to solve the problem. So, two variables require two equations.

Lesson Practice

Complete the activities below to check your understanding of the lesson content.

Questions 1 and 2 refer to the information below.

A holiday nut mix includes peanuts and almonds. Peanuts sell for \$2.25 per pound, and almonds sell for \$4.60 per pound. A bag of holiday mix has 5 pounds of nuts in it and sells for \$15.95.

1. Which linear system of equations can be used to find the number of pounds of peanuts, p, and the number of pounds of almonds, a, that are in the bag of nuts?

   A $\begin{cases} a - p = 5 \\ 4.60a - 2.25p = 15.95 \end{cases}$

   B $\begin{cases} a + p = 15.95 \\ 4.60a + 2.25p = 5 \end{cases}$

   C $\begin{cases} a + p = 5 \\ 4.60a + 2.25p = 15.95 \end{cases}$

   D $\begin{cases} a + p = 5 \\ 2.25a + 4.60p = 15.95 \end{cases}$

   E $\begin{cases} ap = 5 \\ 4.6a + 2.25p = 15.95 \end{cases}$

2. How many pounds of almonds and peanuts are in the bag of holiday mix?

   A 2 pounds of almonds and 3 pounds of peanuts

   B 3 pounds of almonds and 2 pounds of peanuts

   C 1 pound of almonds and 4 pounds of peanuts

   D 4 pounds of almonds and 1 pound of peanuts

   E 2.5 pounds of almonds and 2.5 pounds of peanuts

Questions 3 and 4 refer to the information below.

For three consecutive integers, the sum of the second integer and twice the sum of the first and the third integer is 110.

3. Which of the following equations can be used to determine the numbers?

   A $n + n + 1 + n + 2 = 110$

   B $(n + 1) + 2(n + n + 2) = 110$

   C $(n + 1) + 2(n + n + 3) = 110$

   D $n + 2(n + n + 3) = 110$

   E $n + 2(n + 1 + n + 2) = 110$

4. What are the three consecutive integers?

   A 21, 23, and 25

   B 22, 24, and 26

   C 21, 22, and 23

   D 20, 21, and 22

   E 22, 23, and 24

KEY POINT!

When there are two unknowns in the problem, set up two linear equations.

TEST STRATEGY

Make sure the solution is reasonable for the given context.

Questions 5 and 6 refer to the information below.

A piece of rope that is 28 feet long is cut into two pieces. The longer piece is at least 1 foot longer than twice the length of the shorter piece.

5. Which of the following inequalities can be used to find the length of the shorter piece of rope?

   A $s + (2s + 1) \geq 28$

   B $s - (2s + 1) \geq 28$

   C $28 - s \leq 2s + 1$

   D $28 + s \geq 2s - 1$

   E $28 - s \geq 2s + 1$

6. What is the maximum length of the shorter length of rope?

   A 6 feet

   B 9 feet

   C 12 feet

   D 18 feet

   E 20 feet

Questions 7–9 refer to the information below.

The sum of the angles in any given triangle is 180°. In triangle *ABC*, $m\angle A$ is twice $m\angle B$, and $m\angle C$ is 9 times $m\angle B$.

7. If *x* is the $m\angle B$, which of the following equations can be used to find the measures of the angles?

   A $x + 2x + 9x = 180$

   B $2x + 9x = 180$

   C $x \cdot 2x \cdot 9x = 180$

   D $2x + 2x + 9x = 180$

   E $x - 2x - 9x = 180$

8. In triangle *ABC*, what is the value of *x*?

   A 10

   B 12

   C 15

   D 30

   E 135

9. In triangle *ABC*, what is the measure of each angle?

   A $m\angle A = 10,\ m\angle B = 20,\ m\angle C = 150$

   B $m\angle A = 20,\ m\angle B = 10,\ m\angle C = 150$

   C $m\angle A = 15,\ m\angle B = 30,\ m\angle C = 135$

   D $m\angle A = 30,\ m\angle B = 15,\ m\angle C = 135$

   E $m\angle A = 20,\ m\angle B = 60,\ m\angle C = 100$

See page 91 for answers and help.

Radicals and Radical Equations

Simplifying Radicals

A **radical** is an expression that has a square root symbol. A **radicand** is the value beneath the radical symbol. A numeric radical expression should be simplified using the **Product Property of Square Roots**, which says that $\sqrt{ab} = \sqrt{a} \cdot \sqrt{b}$. When simplifying a radical, look for factors of the number that are perfect squares.

Simplify: $\sqrt{24}$

ANSWER: $\sqrt{24} = \sqrt{4} \cdot \sqrt{6}$
$= 2\sqrt{6}$

SKILLS TIP

Another way to simplify radicals is to rewrite the radicand as a product of prime factors. Each pair of identical factors can be written without the radical symbol because $\sqrt{a \cdot a} = a$. For example, $\sqrt{24} = \sqrt{2 \cdot 2 \cdot 2 \cdot 3} = \sqrt{2 \cdot 2} \cdot \sqrt{2 \cdot 3} = 2\sqrt{6}$

Radicals can be added or subtracted if the radicands are the same. Radicals are combined by adding or subtracting the numbers in front of the radical without changing the radicand. It is a similar thought process to combining like terms in polynomials.

Add: $4\sqrt{2} + 3\sqrt{3} - 2\sqrt{2} + \sqrt{3}$

ANSWER: $4\sqrt{2} + 3\sqrt{3} - 2\sqrt{2} + \sqrt{3} = \left(4\sqrt{2} - 2\sqrt{2}\right) + \left(3\sqrt{3} + \sqrt{3}\right)$
$= 2\sqrt{2} + 4\sqrt{3}$

Many times, some or all of the radicals need to be simplified before they can be combined.

Add: $4\sqrt{12} - 3\sqrt{27} + 5\sqrt{75}$

ANSWER: $4\sqrt{12} - 3\sqrt{27} + 5\sqrt{75} = 4 \cdot \sqrt{4} \cdot \sqrt{3} - 3 \cdot \sqrt{9} \cdot \sqrt{3} + 5 \cdot \sqrt{25} \cdot \sqrt{3}$
$= 4 \cdot 2\sqrt{3} - 3 \cdot 3\sqrt{3} + 5 \cdot 5\sqrt{3}$
$= 8\sqrt{3} - 9\sqrt{3} + 25\sqrt{3}$
$= 24\sqrt{3}$

SKILLS TIP

If a solution to a radical equation is found to be extraneous, then the equation has no solution.

Solving Radical Equations

A **radical equation** is an equation that contains a radical that has a variable in the radicand. To solve a radical equation, isolate the radical first. Then eliminate the radical by squaring it or raising it to a power of 2. Finally, isolate the variable.

Sometimes, a radical equation has an extraneous solution. An **extraneous solution** is simply a solution that, when substituted back into the original equation, does not simplify to a true statement. Every solution of a radical equation must be substituted back into the original equation to see if it is an extraneous solution.

Solve: $\sqrt{5x + 9} - 6 = 6$

ANSWER: $\sqrt{5x + 9} - 6 = 6$

$\sqrt{5x + 9} = 12$ Isolate the radical by adding 6 to both sides.

$5x + 9 = 144$ Square both sides.

$5x = 135$ Subtract 9 from both sides.

$x = 27$ Divide both sides by 5.

Check: $\sqrt{5(27)+9}-6=6$

$\sqrt{135+9}-6=6$

$12-6=6$

$6=6$

The solution is not extraneous, so $x=27$.

Solve: $-1+\sqrt{2x+8}=1$

ANSWER: $-1+\sqrt{2x+8}=1$

$\sqrt{2x+8}=2$ Isolate the radical by adding 1 to both sides.

$2x+8=4$ Square both sides.

$2x=-4$ Subtract 8 from both sides.

$x=-2$ Divide both sides by 2.

Check: $-1+\sqrt{2(-2)+8}=1$

$-1+\sqrt{-4+8}=1$

$-1+\sqrt{4}=1$

$-1+2=1$

$1=1$

The solution is not extraneous. So, $x=-2$.

SKILLS TIP

To eliminate a square root, square both sides of the equation.

Lesson Practice

KEY POINT!

Radicals with the same radicand can be combined by adding or subtracting the numbers in front of the radical while keeping the radicand the same.

TEST STRATEGY

Remember to check your answer to make sure that it is not extraneous by substituting it back into the original equation.

Complete the activities below to check your understanding of the lesson content.

1. Simplify the following expression completely.

   $\left(\sqrt{2}+2\sqrt{3}\right)^2$

   A 14
   B 12
   C $14+4\sqrt{6}$
   D $14+2\sqrt{6}$
   E $6+4\sqrt{6}$

2. Simplify the following expression completely.

   $4\sqrt{18}-2\sqrt{27}+5\sqrt{12}$

   A $16\sqrt{3}$
   B $12\sqrt{2}+4\sqrt{3}$
   C $12\sqrt{2}+16\sqrt{3}$
   D $6\sqrt{2}+10\sqrt{3}$
   E $28\sqrt{3}$

3. Solve the following equation.

   $-2+\sqrt{3x+4}=3$

   A 49
   B 25
   C 7
   D 5
   E no solution

4. Simplify the following expression completely.

   $\sqrt{3}\left(\sqrt{2}+4\sqrt{3}\right)$

   A 15
   B 18
   C $\sqrt{6}+12\sqrt{3}$
   D $\sqrt{6}+12$
   E $\sqrt{6}+36$

5. Solve the following equation.

   $-2+\sqrt{3x-5}=-1$

   A 1
   B 2
   C 16
   D 20
   E no solution

See page 91 for answers and help.

Answer the questions based on the content covered in this unit.

1. Evaluate the expression $3xy - 2z$ if $x = -2$, $y = -1$, and $z = -4$.
   - **A** -14
   - **B** -8
   - **C** 2
   - **D** 14
   - **E** 16

Questions 2 – 4 refer to the information below.

The perimeter of a rectangle is 170 feet. The length, l, of the rectangle is represented by x, and the width, w, is represented by $x + 5$.

2. Which equation represents the area of the rectangle? The formula for the area of a rectangle is $A = lw$.
   - **A** $x(x + 5)$
   - **B** $x + (x + 5)$
   - **C** $2x + (x + 5)$
   - **D** $2x + 2(x + 5)$
   - **E** $(2x + 2) + (x + 5)$

3. Which equation represents the perimeter of the rectangle? The formula for the perimeter of a rectangle is $P = 2l + 2w$.
   - **A** $x(x + 5) = 170$ ft
   - **B** $x + (x + 5) = 170$ ft
   - **C** $2x + (x + 5) = 170$ ft
   - **D** $2x + 2(x + 5) = 170$ ft
   - **E** $(2x + 2) + (x + 5) = 170$ ft

4. What is the area of the rectangle? Use the equation from question 3 to first find the lengths of the sides of the rectangle.
   - **A** 170 ft^2
   - **B** 1,200 ft^2
   - **C** 1,800 ft^2
   - **D** 2,000 ft^2
   - **E** 2,400 ft^2

Questions 5 – 7 refer to the information below.

The diagram below shows a coordinate grid. Use the equation of a line $y = -3x + 6$ to answer the questions.

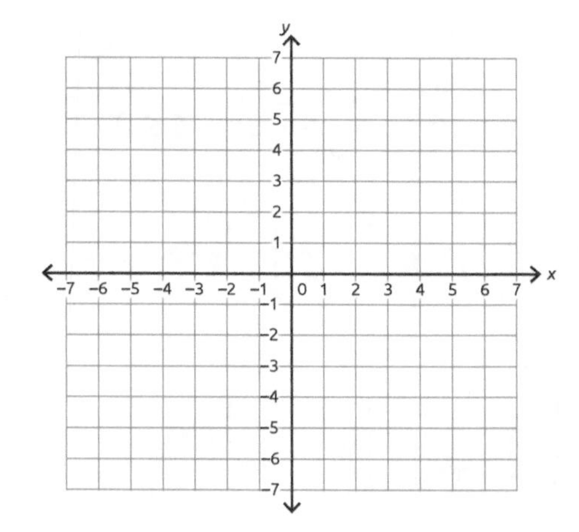

5. What is the slope of the line?
   - **A** $\frac{1}{3}$
   - **B** 3
   - **C** -3
   - **D** -1
   - **E** 1

6. At what point does the graph of the line cross the x-axis?
   - **A** $(2, 0)$
   - **B** $(6, 0)$
   - **C** $(-2, 0)$
   - **D** $(-6, 0)$
   - **E** $(9, 0)$

7. At what point does the graph of the line cross the y-axis?
   - **A** $(0, 3)$
   - **B** $(0, 6)$
   - **C** $(0, -6)$
   - **D** $(0, -3)$
   - **E** $(0, 9)$

8. Simplify the following expression completely.
   $2\sqrt{24}+5\sqrt{32}+4\sqrt{12}$
   - **A** $4\sqrt{6}+20\sqrt{2}+8\sqrt{3}$
   - **B** $20\sqrt{2}+12\sqrt{3}$
   - **C** $24\sqrt{6}+20\sqrt{2}$
   - **D** $4\sqrt{6}+10\sqrt{2}+4\sqrt{3}$
   - **E** $32\sqrt{11}$

9. What is the solution to $-5x-4+9x-4\leq 0$?

A

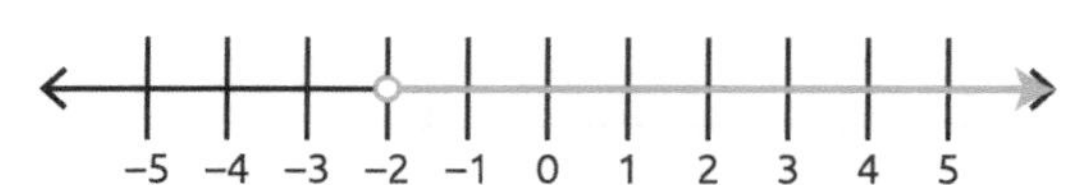

B

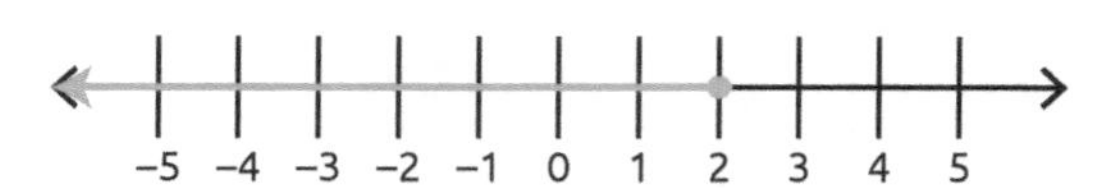

C

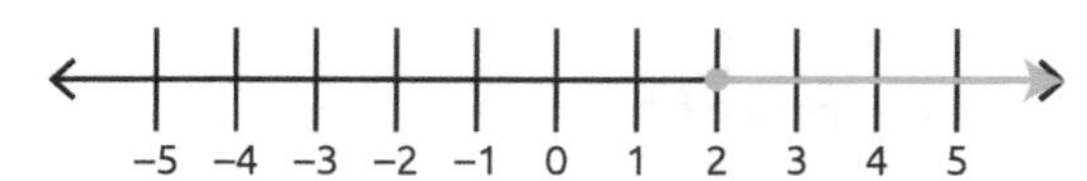

D

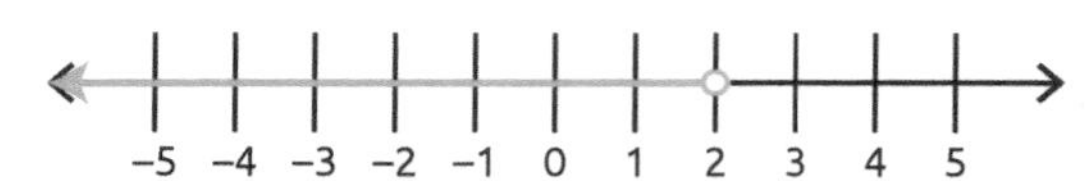

E

Questions 10 and 11 refer to the information below.

10. The formula for the perimeter of a square is $P=4s$. What is the perimeter of the square?
    - **A** 15 cm
    - **B** 30 cm
    - **C** 60 cm
    - **D** 90 cm
    - **E** 225 cm

11. The formula for the area of a square is $A=s^2$. What is the area of the square?
    - **A** 15 cm^2
    - **B** 30 cm^2
    - **C** 60 cm^2
    - **D** 90 cm^2
    - **E** 225 cm^2

12. What is the solution of this linear system of equations?
    $$\begin{cases}3x-4y=2\\5x+3y=13\end{cases}$$
    - **A** (2, 1)
    - **B** (2, 2)
    - **C** (−1, 1)
    - **D** There is no solution.
    - **E** There are infinite solutions.

Questions 13 and 14 refer to the information below.

A large box contains 30 pounds of apples and oranges and is sold for $46.50. Apples sell for $1.35 per pound, and oranges sell for $1.95 per pound.

13. Which linear system of equations can be used to find the number of pounds of apples, *a*, and the number of pounds of oranges, *r*?
    - **A** $\begin{cases}a+r=46.50\\1.35a+1.95r=30\end{cases}$
    - **B** $\begin{cases}a+r=30\\1.35a+1.95r=46.50\end{cases}$
    - **C** $\begin{cases}a+r=30\\1.95a+1.35r=46.50\end{cases}$
    - **D** $\begin{cases}a+r=46.50\\1.95a+1.35r=30\end{cases}$
    - **E** $\begin{cases}a-r=30\\1.35a+1.95r=46.50\end{cases}$

14. How many pounds of apples and how many pounds of oranges are in the box?
 - **A** 18 pounds of apples and 12 pounds of oranges
 - **B** 12 pounds of apples and 18 pounds of oranges
 - **C** 15 pounds of apples and 15 pounds of oranges
 - **D** 10 pounds of apples and 20 pounds of oranges
 - **E** 20 pounds of apples and 10 pounds of oranges

15. To convert the temperature from Fahrenheit to Celsius, use the formula $C = \frac{5}{9}(F - 32)$, where F is degrees Fahrenheit and C is degrees Celsius. To the nearest tenth, what is the temperature in degrees Celsius when the temperature is 18°F?
 - **A** −25.2°C
 - **B** −7.8°C
 - **C** 23°C
 - **D** 7.8°C
 - **E** 41°C

16. Simplify the following expression completely.
 $\sqrt{5}\left(\sqrt{2}+2\sqrt{3}\right)$
 - **A** 35
 - **B** 16
 - **C** $\sqrt{10}+2\sqrt{15}$
 - **D** $5+10\sqrt{3}$
 - **E** $\sqrt{10}+\sqrt{30}$

17. Which of the following graphs represents the graph of the line $y = -\frac{1}{3}x + 1$?

A

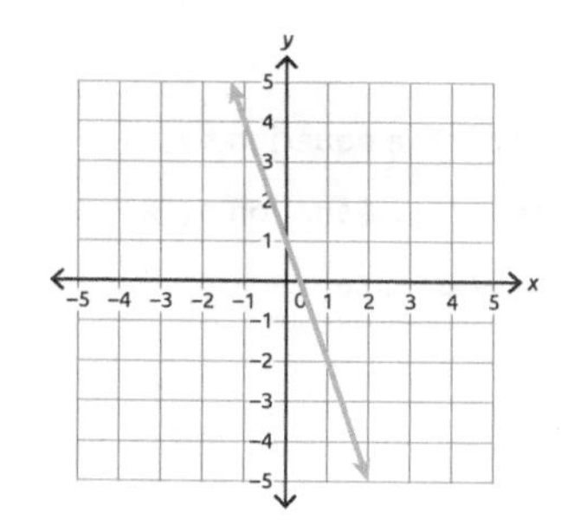

B

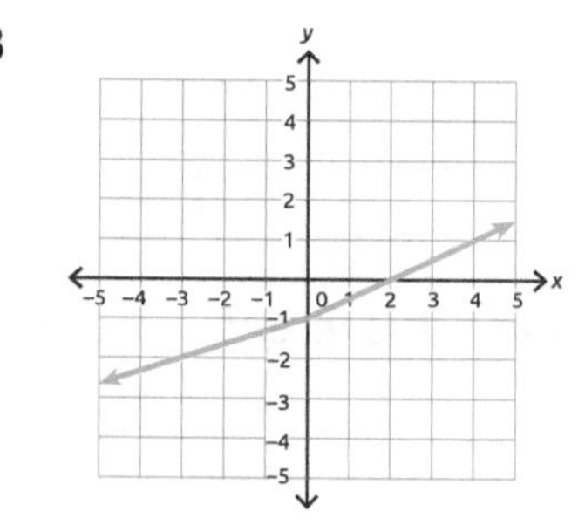

C

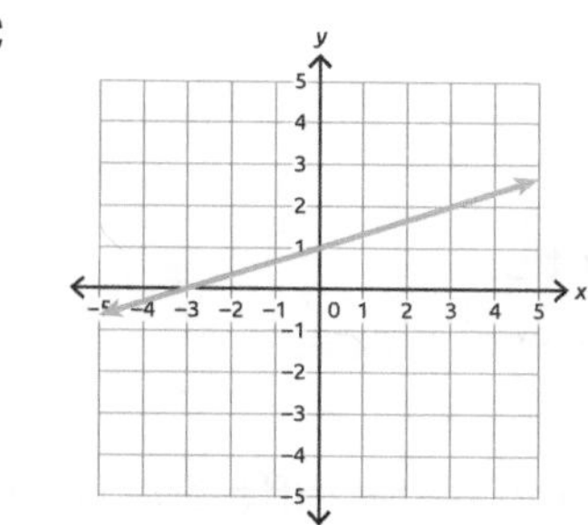

D

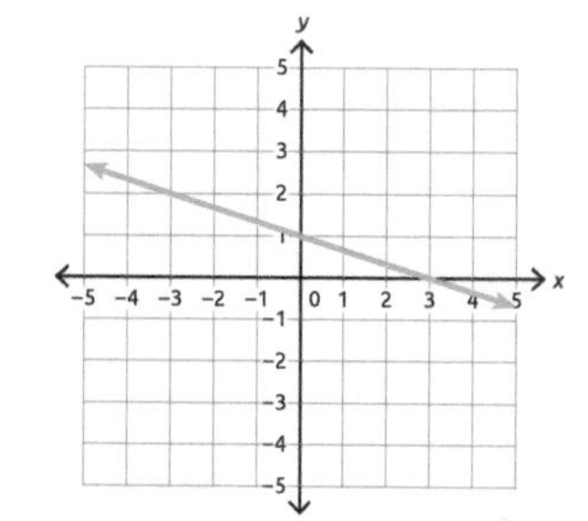

E

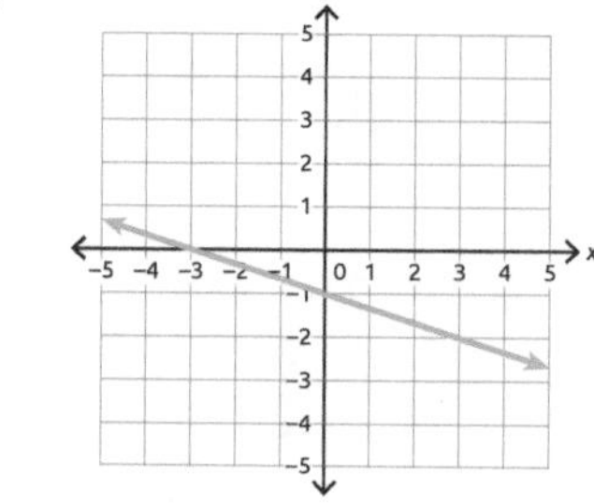

18. Consider the equation $7x + 1 = 3x - 2$. Which of the following steps would be done when solving this equation?

> A. Divide both sides of the equation by 7.
> B. Divide both sides of the equation by 3.
> C. Combine the $+1$ and -2 terms.

- **A** A only
- **B** B only
- **C** C only
- **D** A and B only
- **E** B and C only

19. What is the solution of this linear system of equations?

$$\begin{cases} 12x - 4y = 14 \\ -6x + 2y = -7 \end{cases}$$

- **A** (2, 2)
- **B** (−2, 2)
- **C** (−3, 1)
- **D** There is no solution.
- **E** There are infinite solutions.

20. Which of the following is the difference of the polynomials shown here?

$(6x^2 - x + 4) - (9x^2 - 2x + 5)$

- **A** $-3x^2 + x - 1$
- **B** $-3x^2 - 3x + 9$
- **C** $-3x^2 + x + 9$
- **D** $3x^2 + x - 1$
- **E** $3x^2 - 3x + 9$

21. Which of the following graphs represents the solution to this linear system of equations?

$$\begin{cases} y = \frac{1}{2}x + 3 \\ y = x + 2 \end{cases}$$

A

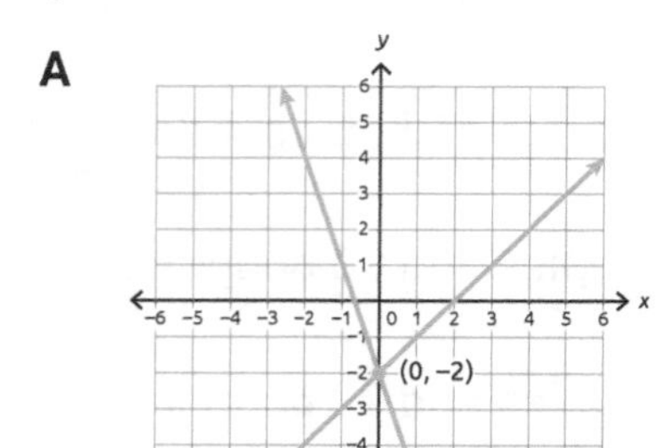

B

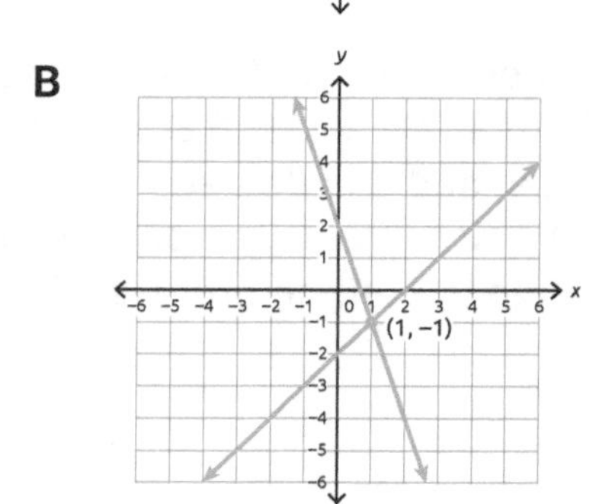

C

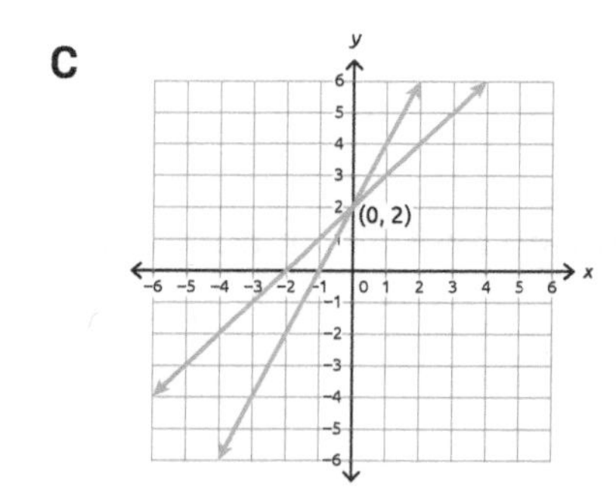

D

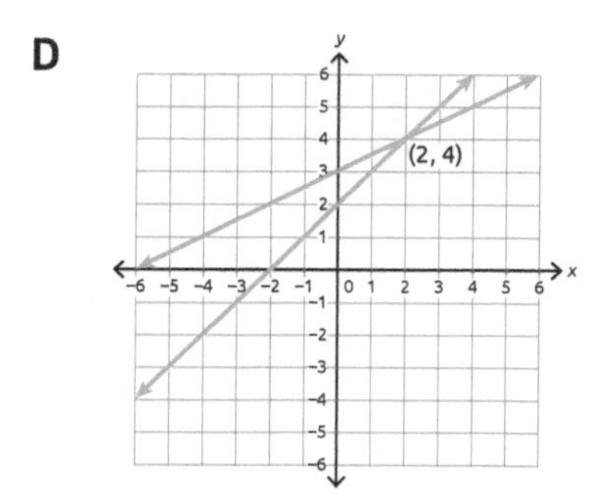

E

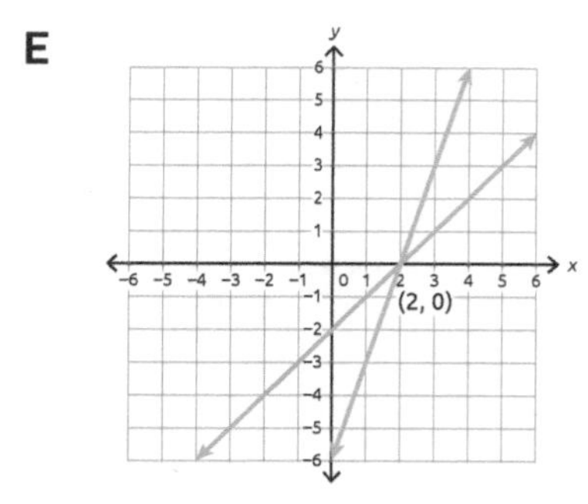

22. The sale price of an item that is marked down 20% can be found using the formula $S = P - 0.2P$, where S is the sale price and P is the original price of the item. What is the sale price of an item that has an original price of $78?
    - **A** $15.60
    - **B** $56.20
    - **C** $62.40
    - **D** $68.00
    - **E** $93.60

23. The circumference of a circle is found using the formula $C = 2\pi r$, where r is the radius of the circle. What is the circumference of the circle whose radius is 8 meters? Use $\pi = 3.14$.
    - **A** 8 m
    - **B** 16 m
    - **C** 25.12 m
    - **D** 50.24 m
    - **E** 200.96 m

Questions 24 and 25 refer to the information below.

Jake is traveling for a business meeting. He receives an amount of money that he can spend on food each day, called a per diem. He also has $45 cash with him. He wants to spend at most a total of $220 for the week (5 days), which includes the $45 cash that he has with him.

24. Which inequalities can be used to determine the total cost per diem Jake receives from his company?
    - **A** $5d + 220 \leq 45$
    - **B** $5d - 220 \leq 45$
    - **C** $5d + 45 \leq 220$
    - **D** $5d - 45 \leq 220$
    - **E** $5d \leq 220 + 45$

25. What is the maximum per diem that Jake receives for one day's meals?
    - **A** $15
    - **B** $20
    - **C** $25
    - **D** $30
    - **E** $35

26. Factor the following expression completely: $x^3 + 6x^2 - 16x$.
    - **A** $x(x^2 + 6x - 16)$
    - **B** $x(x + 8)(x - 2)$
    - **C** $x(x + 2)(x - 8)$
    - **D** $x(x + 4)(x - 4)$
    - **E** $x(x - 4)(x + 4)$

27. On a cross country team, there are $x + 5$ boys and $2x - 1$ girls. Of these, $x + 9$ team members did not qualify for the state meet. Which expression can be used to determine the number of team members who qualified for the state meet?
    - **A** $-2x - 3$ members
    - **B** $2x + 13$ members
    - **C** $2x - 5$ members
    - **D** $4x - 5$ members
    - **E** $4x - 13$ members

Questions 28 and 29 refer to the information below.

Jemma has a pile of nickels and quarters. She has 31 coins in the pile, with a total value of $4.35.

28. Which system of linear equations can be used to find the number of quarters, q, and nickels, n?
    - **A** $\begin{cases} 5n + 25q = 31 \\ 5n + 25q = 435 \end{cases}$
    - **B** $\begin{cases} 0.05n + 0.25q = 31 \\ n + q = 435 \end{cases}$
    - **C** $\begin{cases} 5n + 25q = 31 \\ n + q = 435 \end{cases}$
    - **D** $\begin{cases} n + q = 31 \\ 0.05n + 0.25q = 4.35 \end{cases}$
    - **E** $\begin{cases} n - q = 31 \\ 0.05n + 0.25q = 435 \end{cases}$

29. How many quarters and nickels are in the pile?
    - A 14 quarters and 17 nickels
    - B 15 quarters and 16 nickels
    - C 20 quarters and 11 nickels
    - D 21 quarters and 10 nickels
    - E 22 quarters and 9 nickels

30. The perimeter of a square can be found using the formula $P = 4s$, where s is the length of a side of the square. What is the perimeter of the square shown here?

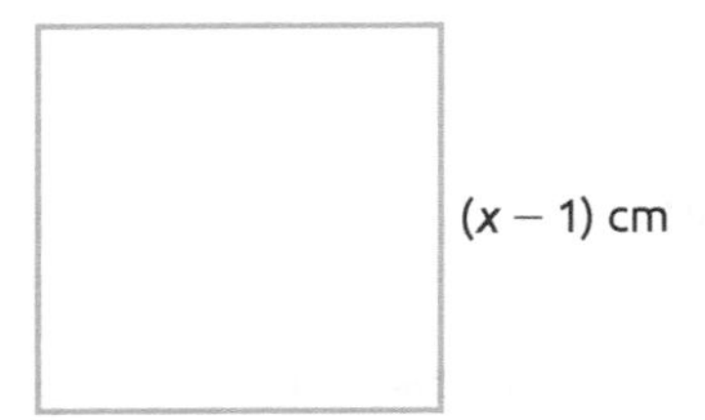

    - A $x^2 - 1$ cm
    - B $x^2 - 2x + 1$ cm
    - C $x^2 - x + 1$ cm
    - D $4x - 4$ cm
    - E $4x - 1$ cm

31. Simplify the following expression completely.

$$\left(\sqrt{3} + 2\sqrt{5}\right)^2$$

    - A 15
    - B 13
    - C $23 + 2\sqrt{15}$
    - D $23 + 4\sqrt{15}$
    - E $17 + 4\sqrt{15}$

Questions 32 and 33 refer to the information below.

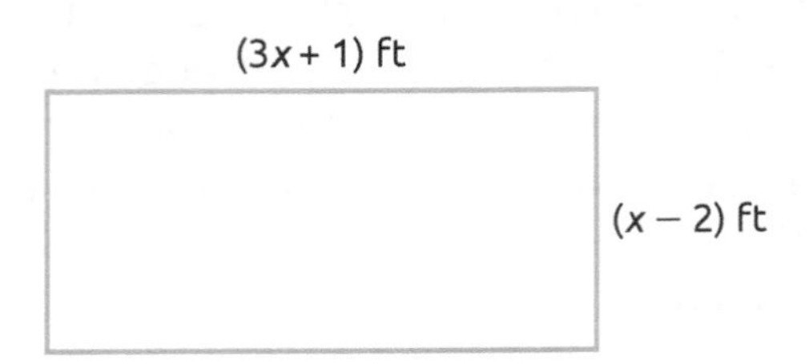

32. The perimeter of a rectangle can be found using the formula $P = 2l + 2w$, where l is the length of the rectangle, and w is the width of the rectangle. Which expression represents the perimeter of the rectangle shown?
    - A $4x - 1$ feet
    - B $8x - 2$ feet
    - C $8x - 1$ feet
    - D $3x^2 - 5x - 2$ feet
    - E $3x^2 - 7x - 2$ feet

33. The area of a rectangle can be found using the formula $A = lw$. Which expression represents the area of the rectangle shown?
    - A $4x - 1$ ft^2
    - B $8x - 2$ ft^2
    - C $8x - 1$ ft^2
    - D $3x^2 - 5x - 2$ ft^2
    - E $3x^2 - 7x - 2$ ft^2

34. What are the solutions to the equation $x^3 + 4x^2 - 21x = 0$?
    - A $\{-7, 3\}$
    - B $\{-3, 7\}$
    - C $\{-7, 0, 3\}$
    - D $\{-3, 0, 7\}$
    - E $\{3, 0, 7\}$

35. Which inequality is represented in the following graph?

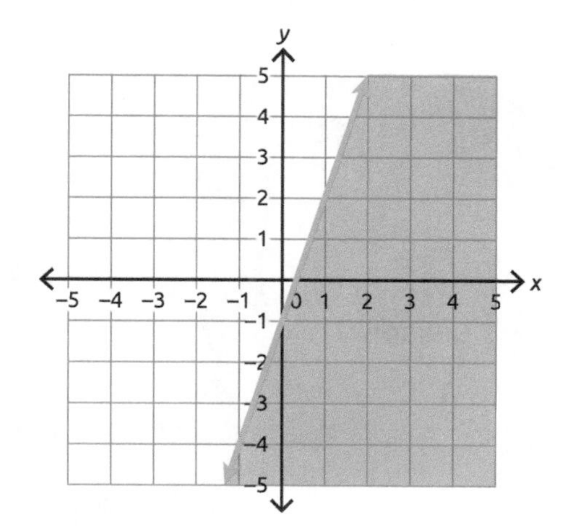

A $y < \frac{1}{3}x - 1$
B $y \leq \frac{1}{3}x - 1$
C $y < 3x - 1$
D $y \geq 3x - 1$
E $y \leq 3x - 1$

Questions 36 and 37 refer to the information below.

For three consecutive even integers, the sum of the first two integers minus three times the third integer equals −40.

36. Which equation can be used to determine the three numbers?
A $n + n + 1 + n + 2 = -40$
B $n + n + 2 + n + 4 = -40$
C $(n + n + 1) - 3(n + 2) = -40$
D $(n + n + 2) - 3(n + 4) = -40$
E $2(n + n + 1) - (n + 2) = -40$

See page 91 for answers.

37. What are the three consecutive integers?
A 30, 31, and 32
B 30, 32, and 34
C 31, 32, and 33
D 31, 33, and 35
E 32, 34, and 36

38. Solve the following equation.
$-3 + \sqrt{6x + 4} = 1$
A 26
B 20
C 2
D 1
E no solution

39. What are the solutions to the quadratic equation $16x^2 + 24x - 7$? Use the quadratic formula.
A $x = \frac{1}{4}, x = -1\frac{3}{4}$
B $x = -\frac{1}{4}, x = 1\frac{3}{4}$
C $x = \frac{1}{2}, x = -3\frac{1}{2}$
D $x = -\frac{1}{2}, x = 3\frac{1}{2}$
E $x = -\frac{1}{4}, x = 3\frac{1}{2}$

40. Which cubic equation has 5 and $3i$ as solutions?
A $x^3 + 5x^2 + 9x + 45$
B $x^3 + 5x^2 - 9x - 45$
C $x^3 - 5x^2 - 9x + 45$
D $x^3 - 5x^2 + 9x - 45$
E $x^3 - 5x^2 - 9x - 45$

Lesson 1

1. B.
2. D. Substitute to get $P = 2(16) + 2(28) = 32 + 56 = 88$ inches.
3. B.
4. C. Use parentheses when substituting −5 for C, and then perform the multiplication before adding 32.
5. A.
6. E.
7. C.
8. D. Substitute 5 for r. Square 5, and then multiply the result by 3.14.
9. B.

Lesson 2

1. E.
2. B. Before combining like terms, change all of the signs of the terms in the second set of parentheses because everything in this set of parentheses is being subtracted. Once all of the signs have been changed, add the like terms.
3. A. Set the expression up as $2(x + 3) + 2(2x - 5)$. Distribute to eliminate the parentheses, and then combine like terms.
4. D. Multiply the expressions $(x + 3) + (2x - 5)$. FOIL the polynomials, and then simplify by combining any like terms.
5. C.
6. D.
7. B. Add the polynomials that represent the number of boys and the number of girls. Next, subtract, from the sum, the polynomial that represents the number of students who left.
8. C.

Lesson 3

1. C. If the equation contains parentheses, the first step is to simplify the parentheses. In this equation, the Distributive Property will help to simplify the multiplication and eliminate the parentheses.
2. E.
3. A.
4. D. The perimeter of a triangle is equal to the sum of its three sides. So, the equation is written as the sum of all three of the sides set equal to 250.
5. B. Solve the equation by combining like terms on the left side to get $3x + 10$. Next, solve for x, and then substitute the value of x into each side length expression to find the length of each side of the triangle.
6. B. Factor the quadratic, and then set both factors equal to 0 according to the Zero-Product Property.
7. D.
8. C.

Lesson 4

1. D. In the equation $y = mx + b$, the m represents the slope. The slope of this line is −2.
2. C. Substitute 0 for y, and then solve for x.
3. D. Substitute 0 for x, and then solve for y.
4. B. The dashed line and the shading above the dashed line indicate a $>$ symbol. The slope is negative because the line is going down from the left to the right.
5. C.

Lesson 5

1. C. Two equations are needed: one for the total value of the tickets and one for the total number of tickets sold. For the total value equation, multiply each adult ticket sold by $6 and each child ticket sold by $4, and set their sum equal to $990. For the total number of tickets sold equation, the sum of the number of adult tickets sold and the number of child tickets sold should be set equal to 185.
2. B. Use substitution or elimination to solve the system of linear equations from question 1.
3. D.
4. A.
5. B.
6. E. "Sum" and "difference" are the key words used to set up the linear equations for the system. "Sum" means add, and "difference" means subtract.
7. D. Elimination can be used to solve the system of linear equations from question 6 since the coefficients of the terms with y are the same in value and have different signs.
8. E. When the top equation is multiplied by 2 and the resulting equation is added to the bottom equation, the result is $0 = 0$, or infinite solutions since the equations are the same line.

Lesson 6

1. **C.** There should be a total cost equation and a total pounds of nuts equation. For the total cost equation, multiply the number of pounds of peanuts by $2.25 and the number of pounds of almonds by $4.60 and set the sum equal to $15.95. For the total pounds of nuts equation, the sum of the number of pounds of peanuts and the number of pounds of almonds should be set equal to 5 pounds.

2. **A.** Use substitution or elimination to solve the system of linear equations from question 1.

3. **B.** The three consecutive integers are represented by the expressions n, $n+1$, and $n+2$. To multiply "twice the sum" use parentheses around the sum of the first integer, n, and the third integer, $n+2$.

4. **C.** Distribute to eliminate the parentheses and the combine like terms. Next, isolate the variable on one side of the equation.

5. **E.** Write an inequality with only one variable. On the left side, the longer piece is 28 feet minus the length of the shorter piece, s. On the right side, the longer piece is two times the shorter piece plus 1, $2s+1$.

6. **B.**

7. **A.** If $m\angle B = x$, then $m\angle A = 2x$, and $m\angle C = 9x$. Also, $m\angle A + m\angle B + m\angle C = 180$.

8. **C.**

9. **D.** Substitute $x = 15$ into $m\angle A = 2x$, $m\angle B = x$, and $m\angle C = 9x$.

Lesson 7

1. **C.** Rewrite this as $(\sqrt{2}+2\sqrt{3})(\sqrt{2}+2\sqrt{3})$ and then FOIL.
   $(\sqrt{2}+2\sqrt{3})(\sqrt{2}+2\sqrt{3}) =$
   $2+2\sqrt{6}+2\sqrt{6}+12 = 14+4\sqrt{6}$

2. **B.** Simplify each square root and then combine the terms with like radicands.

3. **C.** Isolate the radical first, then square both sides of the equation to remove the radical.

4. **D.** Distribute and then simplify.

5. **B.** Isolate the radical first, then square both sides of the equation to remove the radical.

Unit Test

1. D.
2. A.
3. D.
4. C.
5. C.
6. A.
7. B.
8. A.
9. B.
10. C.
11. E.
12. A.
13. B.
14. E.
15. B.
16. C.
17. D.
18. C.
19. E.
20. A.
21. D.
22. C.
23. D.
24. C.
25. E.
26. B.
27. C.
28. D.
29. A.
30. D.
31. D.
32. B.
33. D.
34. C.
35. E.
36. D.
37. B.
38. C.
39. A
40. D

Unit Glossary

- **algebraic expression** – a number sentence involving variables and constants with no equal sign
- **Cartesian coordinate plane** – a coordinate grid separated into four parts or regions by the *x*-axis and the *y*-axis
- **coefficient** – the number in front of a variable
- **constant** – a number; a term without a variable factor
- **equation** – a number sentence that has variables and constants with an equal sign
- **extraneous solution** – a solution of a radical equation that does not check when substituted back into the original equation
- **greatest common factor (GCF)** – the greatest factor that is common to all terms in an expression
- **inequality** – a number sentence with an inequality symbol in place of an equal sign
- **intercept** – the point where a line intersects an axis
- **like terms** – terms with identical variables and powers
- **monomial** – a single-term expression that is made up of a constant, variable(s), or a constant multiplied by a variable(s)
- **order of operations** – the order to follow when simplifying an expression (PEMDAS)
- **polynomial** – a series of different terms that are added or subtracted
- **Product Property of Square Roots** – $\sqrt{ab} = \sqrt{a} \cdot \sqrt{b}$
- **quadratic equation** – a polynomial equation with a degree of 2
- **radical** – an expression with a radical or square root symbol
- **radical equation** – an equation with a radical and a variable in the radicand
- **radicand** – the number or expression under the radical symbol
- **rational expression** – a fraction with variables in the numerator and/or denominator
- **scientific notation** – a system for writing very large or very small numbers using powers of ten
- **slope** – measure of the steepness of a line
- **slope-intercept form** – $y = mx + b$ where m is the slope and b is the y-intercept
- **system of equations** – two or more equations with two or more variables that are common to all of the equations
- **variable** – a letter that represents a number or unknown value
- ***x*-intercept** – the point at which a line intersects the *x*-axis
- ***y*-intercept** – the point at which a line intersects the *y*-axis
- **Zero-Product Property** – if the product of two factors is 0, then one or both factors must be equal to 0

Evaluating Expressions

- Understand variables, constants, and expressions.
- Understand how the order of operations and the Distributive Property can be used on expressions.

Performing Operations on Polynomials and Rational Expressions

- Learn the product and power rules for exponents.
- Understand how to multiply polynomials with more than three terms.
- Know how to square binomials.
- Understand greatest common factor (GCF) and how it is used when factoring.
- Understand how to factor trinomials, including special cases.
- Understand how to simplify rational expressions.

Solving Equations and Inequalities

- Learn how to solve equations, including equations that have decimals and fractions.
- Learn how to solve multi-step inequalities and represent the solution on a number line.
- Learn how to use factoring to solve quadratic equations.

Graphing Equations and Inequalities

- Know how to graph points.
- Learn and understand slope and slope-intercept form of the equation of a line.
- Know how the graphs and slopes of parallel and perpendicular lines relate.
- Learn and understand the slope formula when given two points on a line.
- Learn and understand the point-slope form of the equation of a line.

Solving Systems of Equations

- Know how to use the three different methods of solving a linear system of equations, including a linear system of equations that has fractions or decimals.

Solving Problems Using Expressions, Equations, and Inequalities

- Know how to solve a variety of word problems using expressions, equations, and inequalities.

Radicals and Radical Equations

- Know how to simplify and perform operations with radicals.
- Know how to solve radical equations and check for extraneous solutions.

UNIT 4

Functions

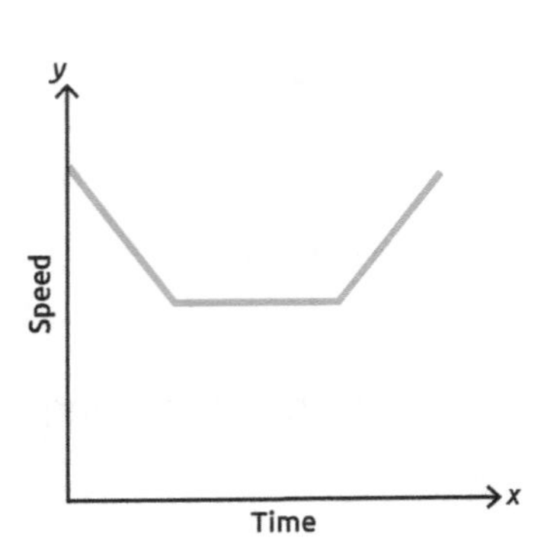

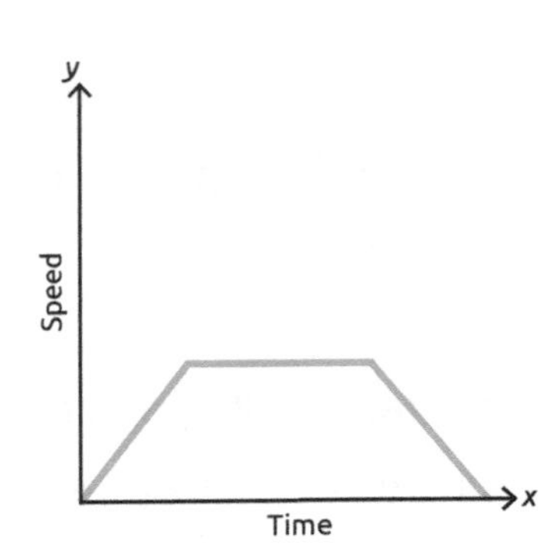

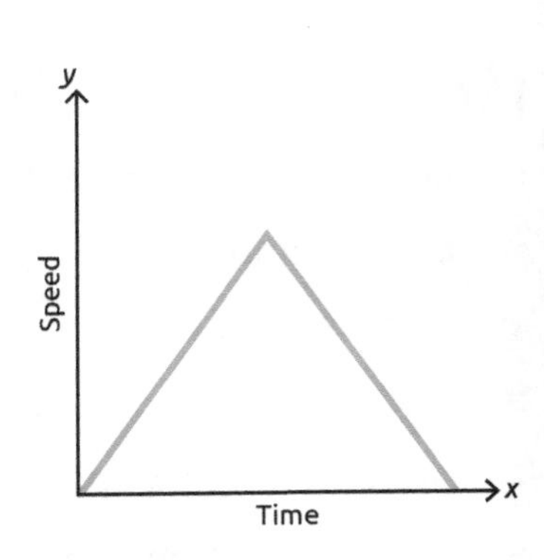

Functions are a very important concept in algebra. Functions allow for predictability because every input has exactly one output. It is important to identify patterns that relate the input to the output. Consider the following scenario:

> Samwell is running on a treadmill for 40 minutes. He gradually increases speed to a certain level. Then, after running at that speed for a time, he gradually decreases his speed until he is finished with his run. Which of the following graphs best represents his run?
>
> **Answer:** The second graph

KEY WORDS

- dependent variable
- domain
- exponential function
- function
- independent variable
- linear function
- quadratic function
- range

Defining, Evaluating, and Comparing Functions

Functions Defined

A **function** is a relation between two sets where each member of the **domain** (inputs) corresponds to exactly one member of the **range** (outputs). So, every input results in or corresponds to exactly one output. The domain is typically represented by the x-values, and the range by the y-values. However, the range is represented by function notation, $f(x)$, which reads "f of x." Domain is the **independent variable**. Range is the **dependent variable** because its value depends on the value of the domain.

A function can be represented by an equation, a table of values, a graph, or a mapping diagram.

<table>
<tr><th>Equation</th><th>Table</th><th>Graph</th><th>Mapping</th></tr>
<tr><td>$f(x) = x + 1$</td><td><table><tr><th>x</th><th>f(x)</th></tr><tr><td>1</td><td>2</td></tr><tr><td>2</td><td>3</td></tr><tr><td>3</td><td>4</td></tr></table></td><td>(2, 3)
(0, 1)
(−1, 0)</td><td>2 → 3
0 → 1
−1 → 0</td></tr>
<tr><td>Write an equation using function notation.</td><td>Choose any value of x, and then find the corresponding value of f(x).</td><td>The function can be represented on a graph.</td><td>A mapping diagram shows the pairings of domain with range.</td></tr>
</table>

Evaluating Functions

A function is evaluated for any value in the domain. The value simply gets plugged into the function to find the corresponding value in the range. The notation is a little different for functions. To find the value in the range when $x = 2$ for example, find $f(2)$. Consider the following example to help you understand how to evaluate a function.

If $g(x) = 3x - 2$ and $h(x) = x^2 + 6$, find $g(-4)$ and $h(-2)$.

ANSWER: $g(-4) = 3(-4) - 2 = -12 - 2 = -14$

So, $g(-4) = -14$.

$h(-2) = (-2)^2 + 6 = 4 + 6 = 10$

So, $h(-2) = 10$.

SKILLS TIP

Often, a machine is used to illustrate a function, where an equation is the machine. Every value input into the function machine results in exactly one output value.

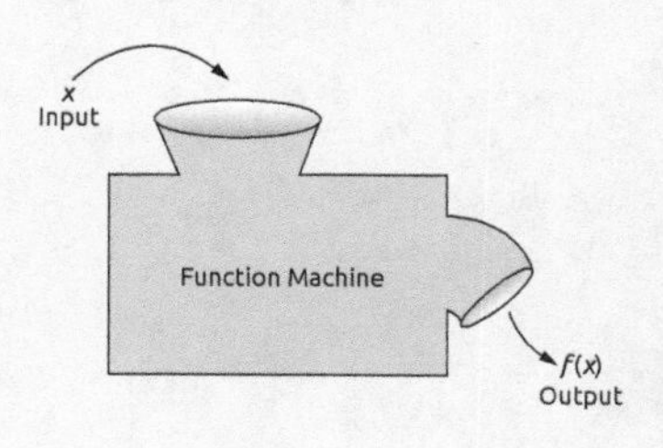

SKILLS TIP

Regardless of how a function is represented, it is only a function if each member of the domain is paired with exactly one member of the range.

SKILLS TIP

When substituting a value into an expression, use parentheses around the value.

Comparing Inverse Functions

If a function is represented by the ordered pair (a, b), then the inverse function is the ordered pair (b, a). The inverse of $f(x)$ is denoted $f^{-1}(x)$.

Find the inverse of the function $f(x) = -2x - 3$.

ANSWER: Follow four steps to find the inverse of a function in equation form.

$f(x) = -2x - 3$	
$y = -2x - 3$	Replace $f(x)$ with y.
$x = -2y - 3$	Interchange the variables.
$x + 3 = -2y$	Solve for y.
$\dfrac{x+3}{-2} = y$	
$f^{-1}(x) = \dfrac{x+3}{-2}$	Inverse function

When the graph of a function is reflected over $y = x$, the result is the graph of the inverse of the function.

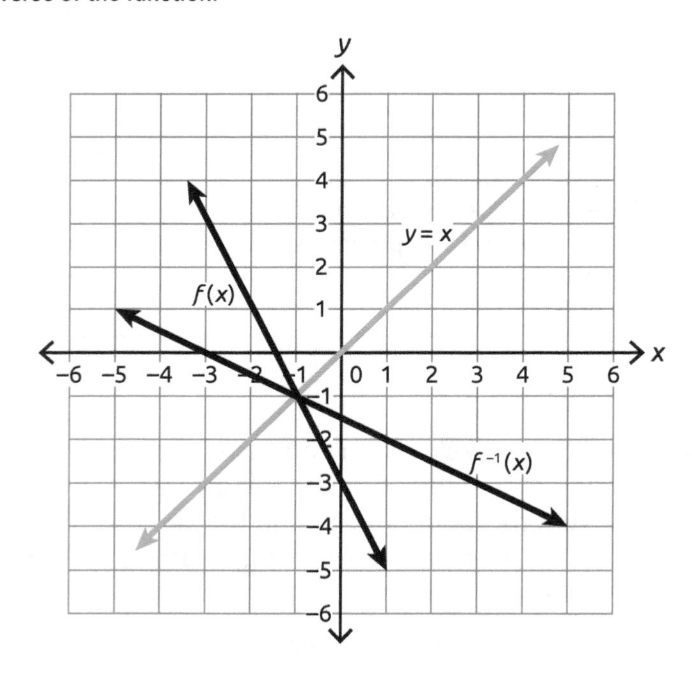

Complete the activities below to check your understanding of the lesson content.

1. If $f(x) = x^2 - 1$, what is $f(-5)$?
   A −26
   B −24
   C 9
   D 24
   E 26

2. The number of boys, b, in the class varies directly as the number of girls, g, and inversely as the number of teachers, t. Which of the following equations represents the relationship?
   A $b = gt$
   B $b = g + \frac{1}{t}$
   C $b = \frac{t}{g}$
   D $b = \frac{g}{t}$
   E $b = \frac{1}{gt}$

3. The functions $g(x)$ and $h(x)$ are inverse functions. They appear as mirror images when reflected over which line?
   A $y = -x$
   B $y = x$
   C y-axis
   D x-axis
   E $x = 0$

4. If $j(x) = 5x + 7$, what is $j(-1)$?
   A 12
   B 2
   C −2
   D −12
   E −35

KEY POINT!

In a function, each member of the domain corresponds to exactly one element of the range.

TEST STRATEGY

Pay close attention to the rules for signed numbers when following order of operations to simplify an expression.

Lesson Practice

5. Which of the following tables does not represent a function?

A

x	$f(x)$
1	1
2	1
3	1
4	1
5	1

B

x	$f(x)$
−1	−1
−2	−2
−3	−3
3	3
2	2

C

x	$f(x)$
1	1
2	2
3	3
2	4
1	5

D

x	$f(x)$
−1	1
−2	2
−3	3
−4	4
−5	5

E

x	$f(x)$
−1	2
2	4
−3	6
4	4
−5	2

See page 106 for answers and help.

Compare Linear, Quadratic, and Exponential Functions

Various types of functions share features, including general equations and the shape of the graph of the function. Three of the common types of functions include **linear**, **quadratic**, and **exponential functions**. Consider the following table, which describes the key features of each function type.

Name of Function	General Shape of Graph	Key Features of the Equation	Example
linear	line	First-degree equation: $f(x) = 4x - 1$	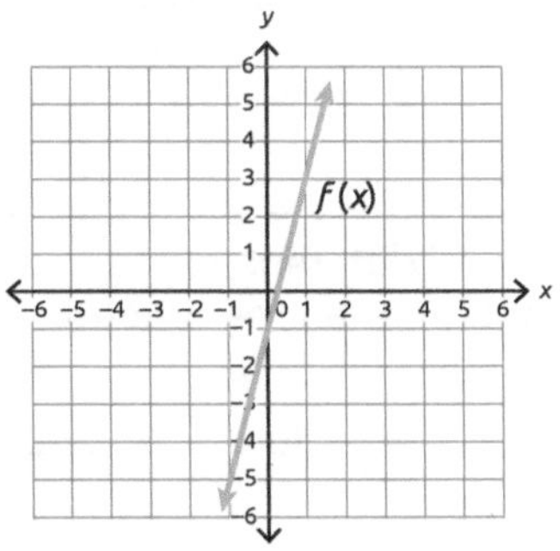
quadratic	parabola	Second-degree equation: $g(x) = x^2 + 2x - 5$	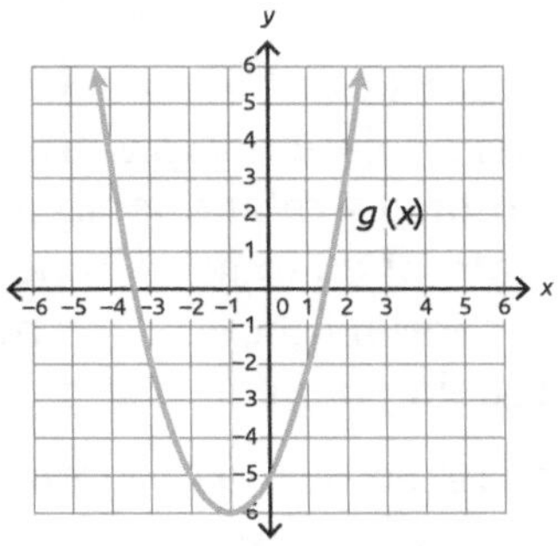
exponential	a rapidly increasing or decreasing curve	An equation with a variable in the exponent: $h(x) = 2^x + 1$	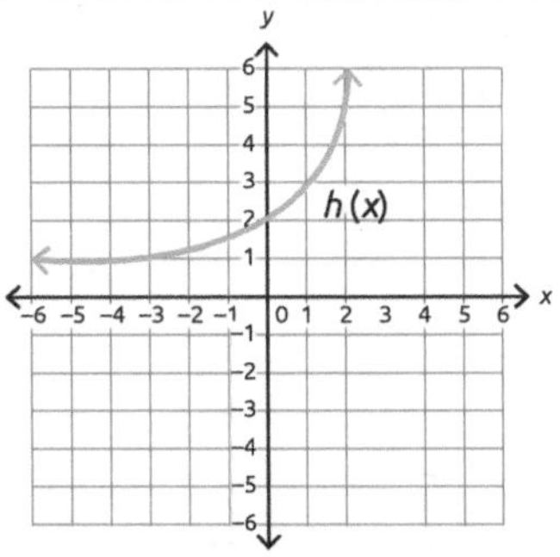

SKILLS TIP

The degree of an equation is equal to the greatest exponent of any term found in the equation.

Identify the type of function for each of the following representations. Explain your reasoning.

a. $g(x) = 3^x - 1$

b.

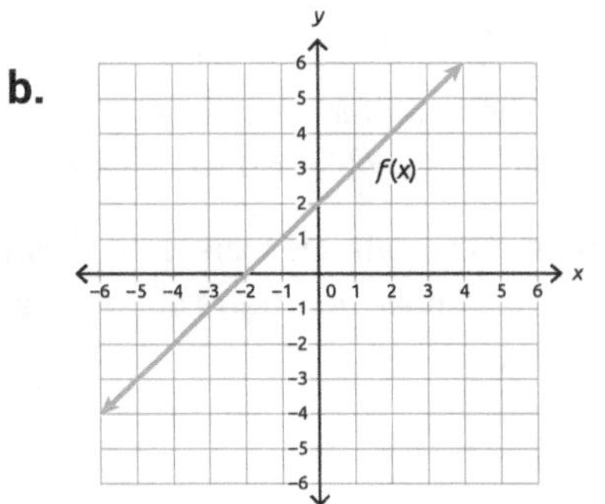

ANSWER: a. $g(x)$ is an exponential function because it has a variable as an exponent.

b. $f(x)$ is linear because the graph forms a straight line.

When looking at a table of values, the function type can be determined by examining the first or second differences. Look at the difference from one *y*-value to the next.

<table>
<tr><th>Linear</th><th>Quadratic</th><th>Exponential</th></tr>
<tr>
<td>
<table>
<tr><td>x</td><td>−3</td><td>−2</td><td>−1</td><td>0</td><td>1</td><td>2</td><td>3</td></tr>
<tr><td>y</td><td>−8</td><td>−5</td><td>−2</td><td>1</td><td>4</td><td>7</td><td>10</td></tr>
</table>
3 3 3 3 3 3
</td>
<td>
<table>
<tr><td>x</td><td>−3</td><td>−2</td><td>−1</td><td>0</td><td>1</td><td>2</td><td>3</td></tr>
<tr><td>y</td><td>9</td><td>4</td><td>1</td><td>0</td><td>1</td><td>4</td><td>9</td></tr>
</table>
−5 −3 −1 1 3 5
2 2 2 2 2
</td>
<td>
<table>
<tr><td>x</td><td>−3</td><td>−2</td><td>−1</td><td>0</td><td>1</td><td>2</td><td>3</td></tr>
<tr><td>y</td><td>$\frac{1}{8}$</td><td>$\frac{1}{4}$</td><td>$\frac{1}{2}$</td><td>1</td><td>2</td><td>4</td><td>8</td></tr>
</table>
$\frac{1}{8}$ $\frac{1}{4}$ $\frac{1}{2}$ 1 2 4
</td>
</tr>
<tr>
<td>If the 1st differences are constant, then the function is linear.</td>
<td>If the 2nd differences are constant, then the function is quadratic.</td>
<td>If the 1st differences follow a pattern similar to the y-values, then the function is exponential.</td>
</tr>
</table>

Use Functions to Model Relationships

Functions can be used to model relationships. A linear function can be used to model a relationship with a rate of change that is constant. A quadratic function can model a relationship with a maximum or minimum value: a point where the rate of change switches from increasing to decreasing or vice versa. An exponential function can model a relationship where the rate of change is either increasing or decreasing at an ever-changing rate.

SKILLS TIP

Various real world activities are modeled by the same function types. Things that increase or decrease at a regular rate are linear. Quadratic functions often model things that are launched into the air. Exponential functions model things like population growth or compound interest. The rate is increasing or decreasing.

Paco hiked over a mountain peak to the other side of a mountain. Which of the following graphs most likely models Paco's altitude as he climbs over the mountain? What type of function best fits this type of graph? Explain your reasoning.

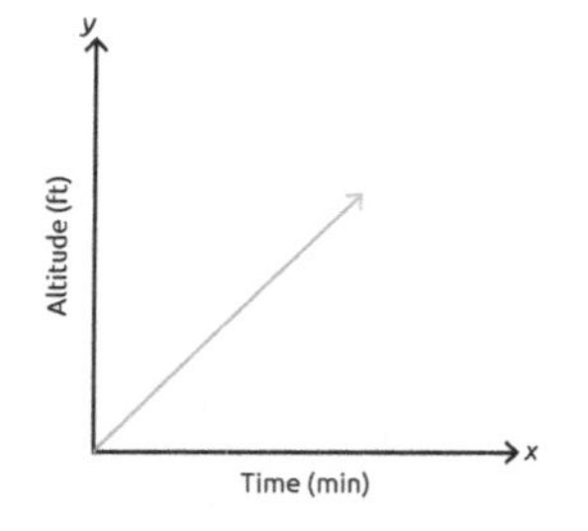

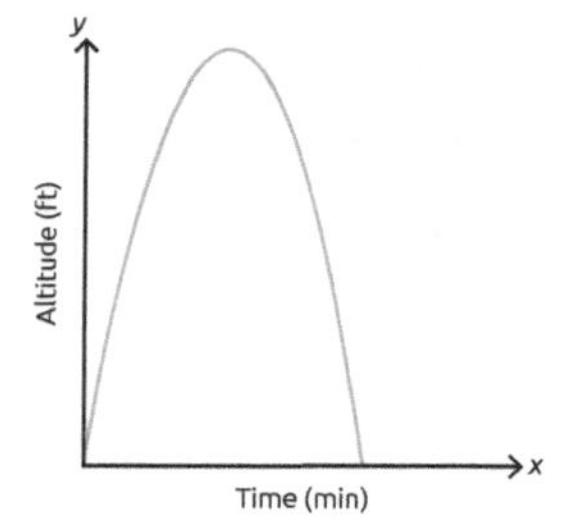

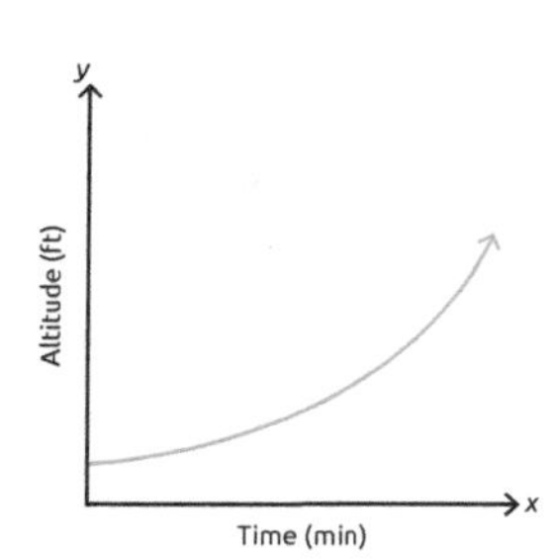

ANSWER: The middle graph models Paco's altitude because it increases, reaches a maximum point, and then decreases. This graph is a parabola, which models a quadratic function.

The graph on the left shows altitude and time increasing at a steady rate. This graph represents a climber who is steadily climbing higher over a period of time.

The graph on the right shows altitude increasing slowly at first and then increasing more rapidly. This graph represents a climber who starts out gradually, slowly climbing, and then begins to climb at a faster rate.

Complete the activities below to check your understanding of the lesson content.

Questions 1 and 2 refer to the graphs on the following grid.

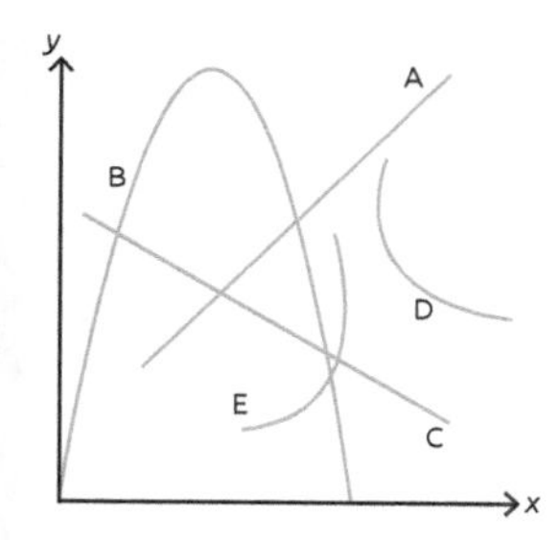

1. According to the shape and the direction of the graph, which figure could correspond to the function $f(x) = -0.5x + 8$?

A A
B B
C C
D D
E E

2. Which graph models a quadratic function?

A A
B B
C C
D D
E E

Questions 3–5 refer to the following choices.

L **linear function**
Q **quadratic function**
E **exponential function**

3. Which types of functions could model the equation $f(x) = 3x^2 + 4x - 1$?

A L only
B Q only
C E only
D Q and E
E None of the functions listed

4. The following graph represents the data in a science experiment modeling the growth of bacteria. Which type of function should be used to model the data?

A L only
B Q only
C E only
D L and Q only
E None of the functions listed

5. Which of the function types does this table represent?

x	$f(x)$
1	−5
2	−1
3	3
4	7
5	11

A L only
B Q only
C E only
D L and Q only
E Q and E only

See page 106 for answers and help.

TEST STRATEGY

Pay close attention to clues in the context of questions and answer choices. For example, linear has the word "line" in it. Linear functions form a line.

Answer the questions based on the content covered in this unit.

1. If $g(x) = 3x^2 + 4$, what is $g(-2)$?
   A 16
   B 12
   C 10
   D −2
   E −8

Questions 2–4 refer to the information below.

L **linear function**
Q **quadratic function**
E **exponential function**

2. Which function is modeled by the equation $f(x) = 3x + 4$?
   A L only
   B Q only
   C E only
   D Q and E
   E None of the functions listed

3. The following graph represents a model of the flight of a ball thrown in the air. Which type of function should be used to model the data?

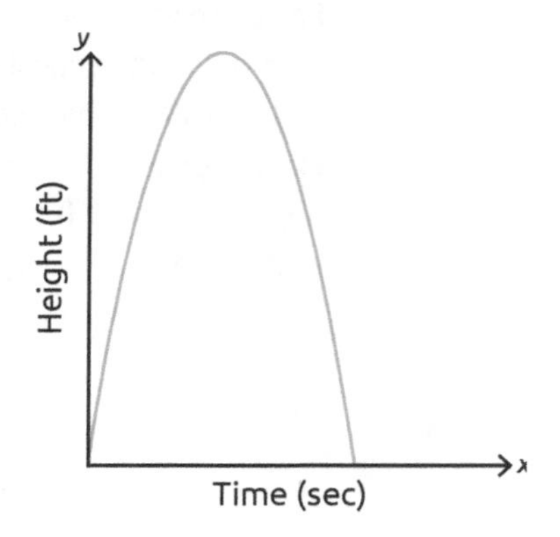

   A L only
   B Q only
   C E only
   D L and Q only
   E None of the functions listed

4. Which of the function types does this table represent?

x	$f(x)$
1	2
2	4
3	8
4	16
5	32

   A L only
   B Q only
   C E only
   D L and Q only
   E Q and E only

5. If the graphs of the functions $f(x)$ and $j(x)$ are mirror images when reflected over the line $y = x$, what is the relationship of these two functions?
   A They are opposite functions.
   B They are negative functions.
   C They are similar functions.
   D They are the same function.
   E They are inverse functions.

6. Which of the following tables represents a function?

A

x	f(x)
1	1
2	1
3	1
4	1
4	1

B

x	f(x)
−1	−1
−2	−2
−3	−3
3	3
2	2

C

x	f(x)
1	1
2	2
3	3
2	4
1	5

D

x	f(x)
0	1
0	2
0	3
0	4
0	5

E

x	f(x)
−1	2
2	4
−1	6
4	4
−1	2

7. The number of red marbles, r, in a bag varies directly as the number of green marbles, g, and inversely as the number of blue marbles, b. Which of the following equations represents the relationship?

A $r=\frac{g}{b}$

B $r=\frac{b}{g}$

C $r=g+\frac{1}{b}$

D $r=gbt$

E $r=\frac{1}{gb}$

Questions 8 and 9 refer to the information below.

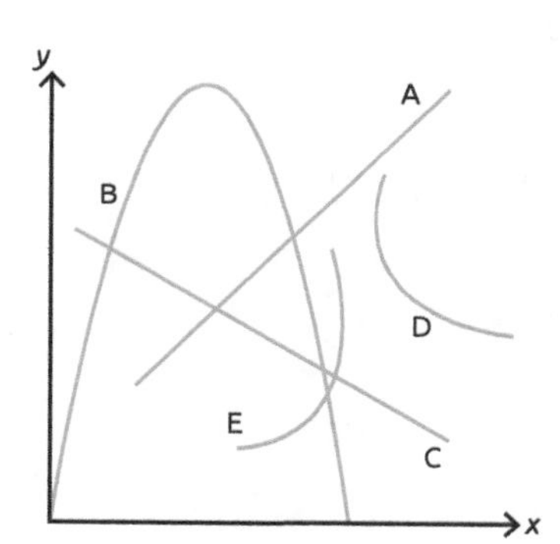

8. According to the shape and the direction of the graph, which figure could correspond to the function $f(x) = 2^x + 1$?

A A

B B

C C

D D

E E

9. Which graph models an increasing linear function?

A A

B B

C C

D D

E E

10. Which of the following mapping diagrams represents a function?

A

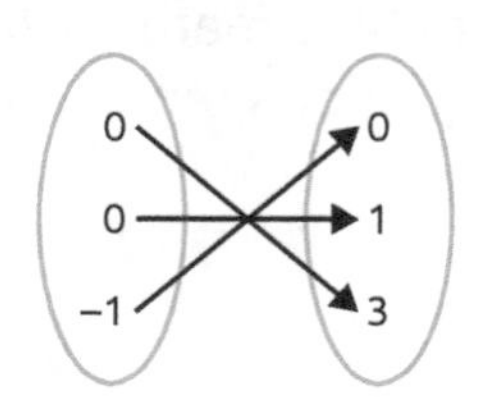

D

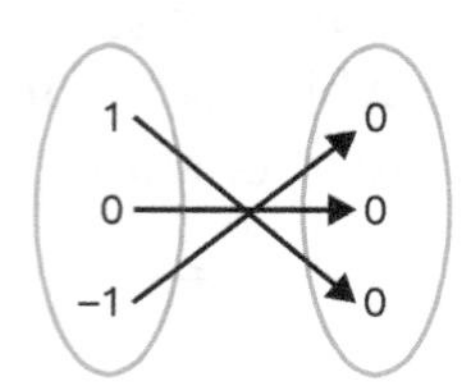

B

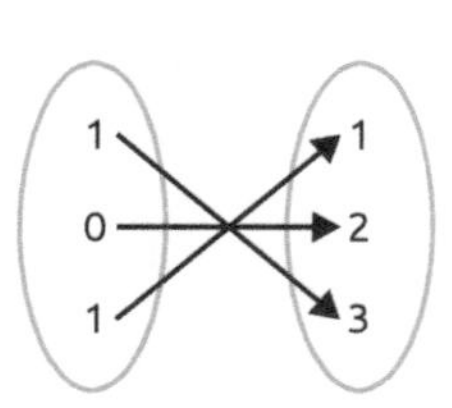

E

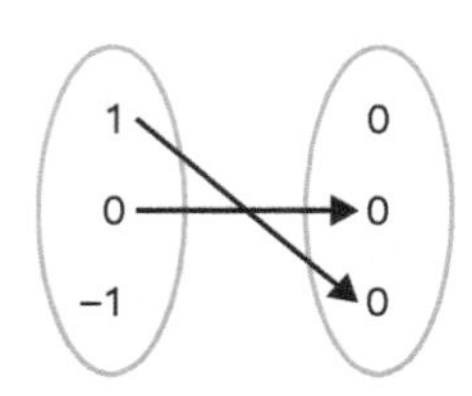

C

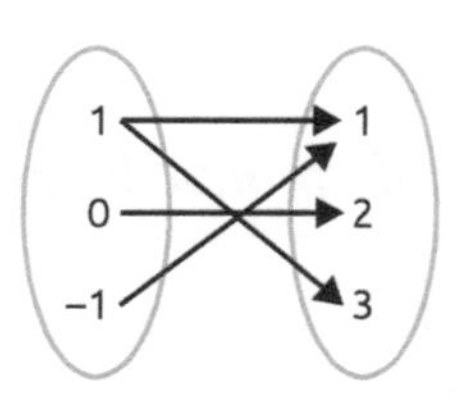

11. If $b(x) = -3x - 5$, what is $b(-6)$?

A 23

B 13

C 8

D −13

E −23

12. The graph below shows the number of bald eagles in New York from 1967–2007. Which type of function best models the population growth shown in the graph?

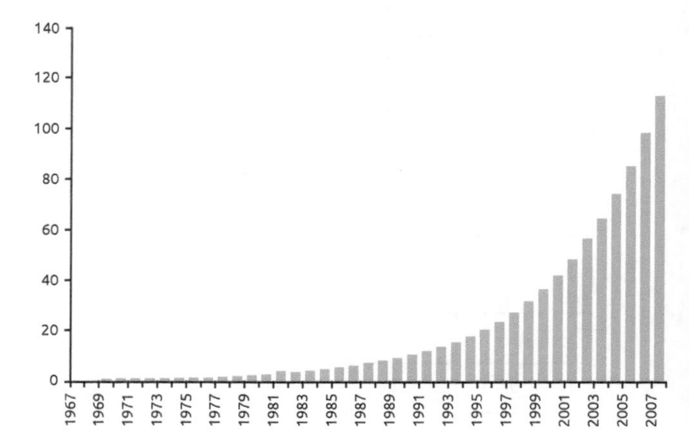

A linear

B quadratic

C exponential

D linear and exponential

E linear and quadratic

13. If $f(x) = 2x + 6$, which of the following is $f^{-1}(x)$?

A $f^{-1}(x) = -2x - 6$

B $f^{-1}(x) = x + 3$

C $f^{-1}(x) = \frac{1}{3}x - 3$

D $f^{-1}(x) = \frac{1}{2}x + 3$

E $f^{-1}(x) = \frac{1}{2}x - 3$

14. Which of the following graphs shows a relationship between x and y where an increase in the value of x always results in an increase in y?

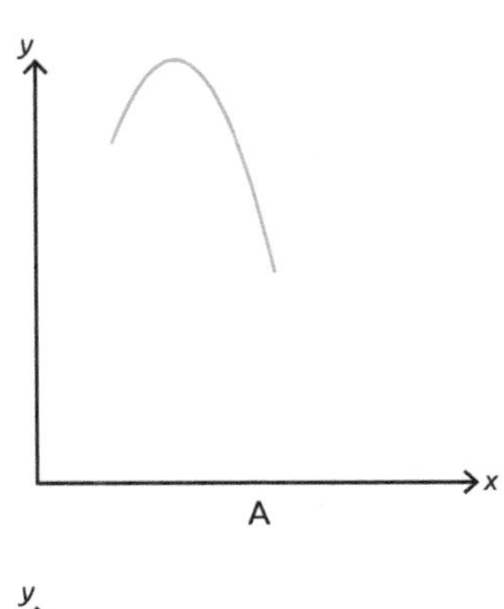

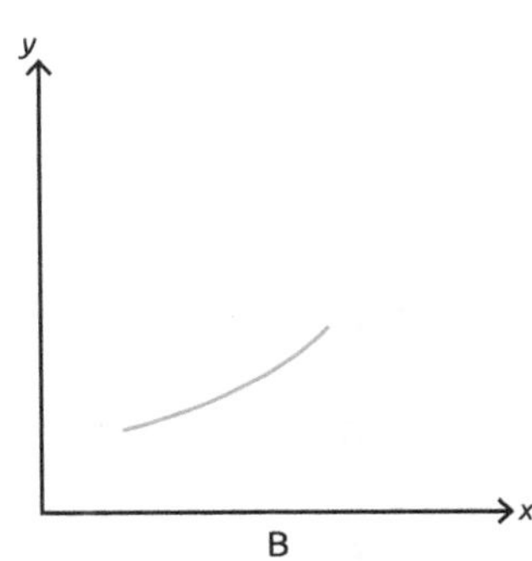

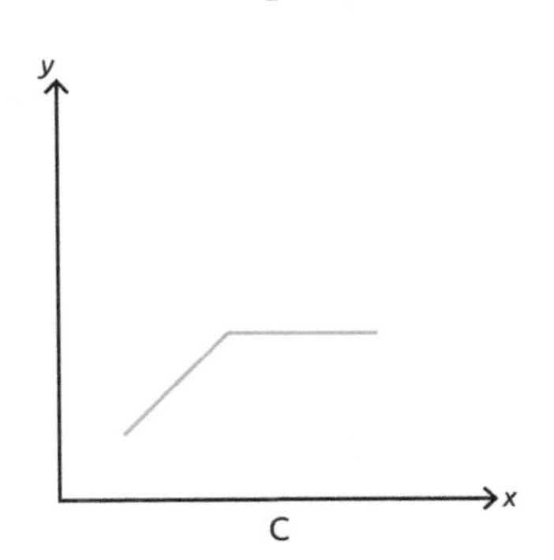

A A only
B B only
C C only
D A and B only
E B and C only

15. Which of the following graphs shows a function and its inverse?

A

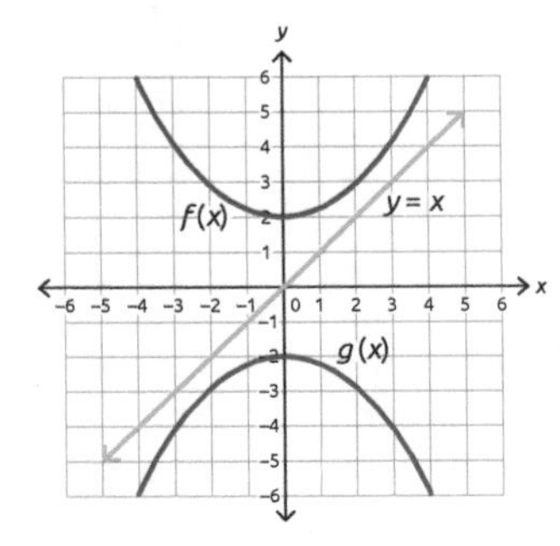

B

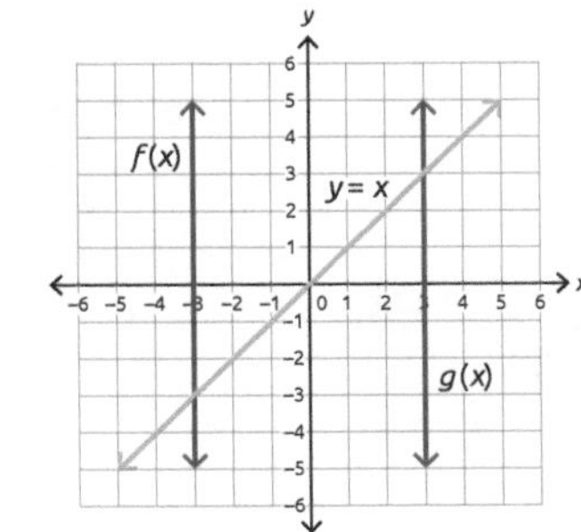

C

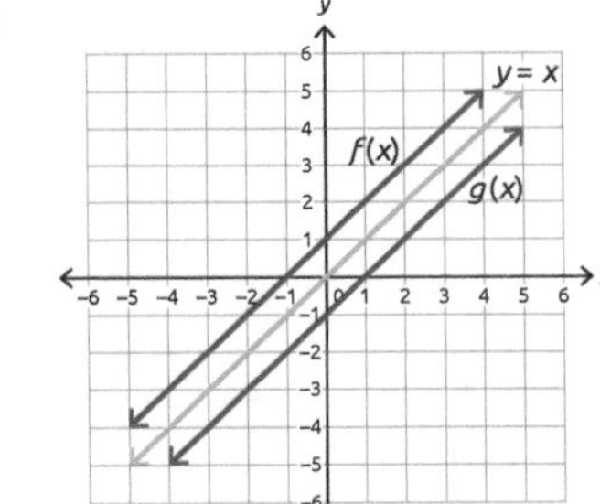

D

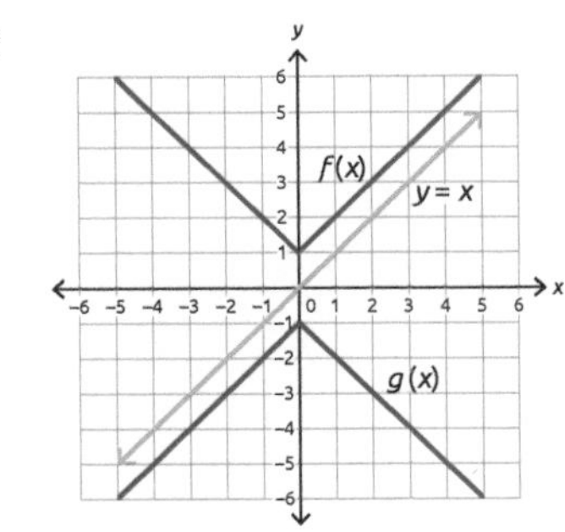

E

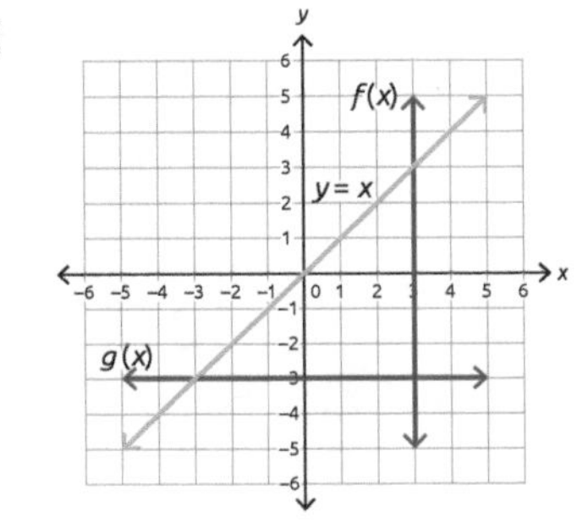

See page 106 for answers.

Unit Answer Key

Lesson 1

1. **D.** Use parentheses when substituting −5 into the function. $f(-5) = (-5)^2 - 1 = 25 - 1 = 24$

2. D.

3. **B.** The graphs of inverse functions are reflections of each other over the line $y = x$.

4. **B.** Use parentheses when substituting −1 into the function. $j(-1) = 5(-1) + 7 = -5 + 7 = 2$

5. **C.** Each number in the domain must only be paired with one number in the range. If a number in the domain pairs with more than one number in the range, like 2 and 1, then it is not a function.

Lesson 2

1. C.

2. **B.** The function is degree 2, which is a quadratic function.

3. **B.** A quadratic function forms a parabola.

4. **C.** The graph represents the curve of an exponential function.

5. **A.** The 1st differences are constant, so the function is linear.

Unit Test

1. A.
2. A.
3. B.
4. C.
5. E.
6. B.
7. A.
8. E.
9. A.
10. D.
11. B.
12. C.
13. E.
14. B.
15. C.

Unit Glossary

- **dependent variable** – output of a function; range
- **domain** – inputs of a function; independent variable; usually *x*
- **exponential function** – a function whose equation has a variable in the exponent and graph is increasing or decreasing at an increasing rate
- **function** – a relation in which each member of the domain corresponds to exactly one member of the range
- **independent variable** – input of a function; domain
- **linear function** – a function whose equation is degree 1, rate of change is constant, and graph is a line
- **quadratic function** – a function whose equation is degree 2 and graph is a parabola
- **range** – outputs of a function; dependent variable; represented by $f(x)$

Study More!

Define, Evaluate, and Compare Functions

- Understand how to evaluate functions for non-numeric values.
- Understand how to perform the composition of functions and use them to determine if functions are inverses of each other.

Use Functions to Model Relationships

- Understand key aspects of a linear function like *x*- and *y*-intercepts and the slope of the function.
- Understand key points in a quadratic function like maximum point or minimum point and the zeros of the function.
- Understand key points in an exponential function, like *x*- and *y*-intercepts, and asymptotes.

UNIT 5

Geometry

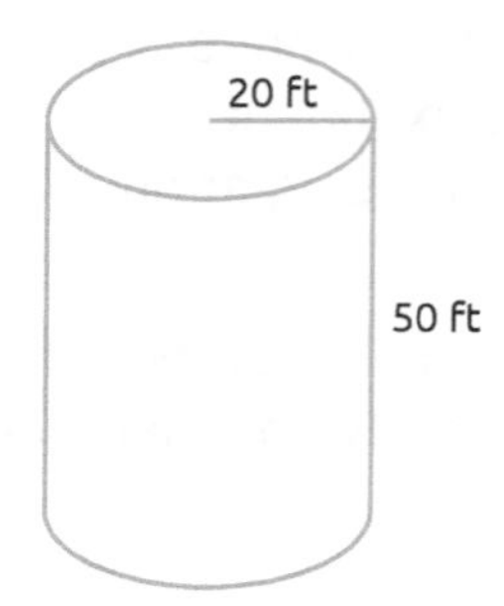

Geometry is a study of spatial relationships. It includes all sorts of 2- and 3-dimensional shapes. It uses words, drawings, and formulas to describe these various relationships. Consider the following scenario:

> Steve has a cylindrical tank that holds the milk from his dairy cows. Each day on average, he is able to fill three-fourths of the tank. About how many cubic feet of milk does he collect on an average day?
>
> **Answer:** about 47,100 ft^3

KEY WORDS

- acute angle
- angle
- complementary angles
- dilation
- edge
- endpoint
- face
- hypotenuse
- image
- lateral area
- line
- line segment
- obtuse angle
- plane
- point
- preimage
- Pythagorean Theorem
- ray
- reflection
- right angle
- rotation
- scale factor
- supplementary angles
- surface area
- transformation
- translation
- vertex
- volume

Shapes Defined

The basic building blocks of geometry are the point, line, and plane. A **point** names a location but has no size and is represented by a dot. A **line** is a straight path made up of an infinite number of points. It has no thickness and extends in both directions indefinitely. A **plane** is a flat surface with no thickness and extends forever in each direction.

Lines and points compose figures called rays and segments. These are defined and illustrated in the table below.

Term	Definition	Drawing
line segment	A part of a line that consists of two endpoints and all of the points between them. Denoted ($\overline{AB}$).	A B
endpoint	Points at each end of a line segment or one end of a ray. Points *A* and *B* are endpoints.	A B
ray	A part of a line with one endpoint and extending forever in one direction. Denoted ($\overrightarrow{AB}$).	A B
angle	Two rays sharing a common endpoint. Denoted $\angle ABC$ or $\angle B$.	A B C

SKILLS TIP

A point is typically named by a single letter. A line is named by two letters—one at each end of the line. A plane can be named by any three points that lie on the plane.

Angles

An angle is measured by degrees. Angles can be classified by their degree measurement: acute, right, or obtuse.

Acute Angle	Right Angle	Obtuse Angle
$0° \leq A < 90°$	$A = 90°$	$90° < A \leq 180°$
A B C	A B C	A B C

SKILLS TIP

A straight angle is 180° and forms a straight line.

What type of angle is shown? Explain your reasoning.

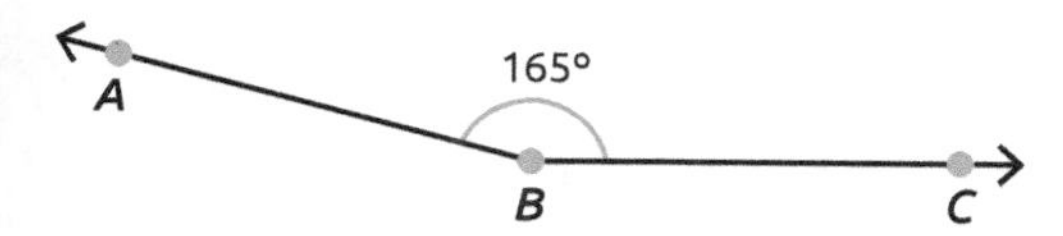

ANSWER: Obtuse; the angle is greater than 90° and less than 180°.

Knowing Shapes and Their Attributes

KEY POINT!

A figure is 2-dimensional, and a solid is 3-dimensional.

Three-Dimensional Solids

A 3-dimensional figure is a solid that has height, width, and depth. A solid is formed by edges, faces, and vertices. A **face** is each flat surface of a solid. An **edge** is the segment formed where two faces intersect. A **vertex** is a point formed by the intersection of three or more faces.

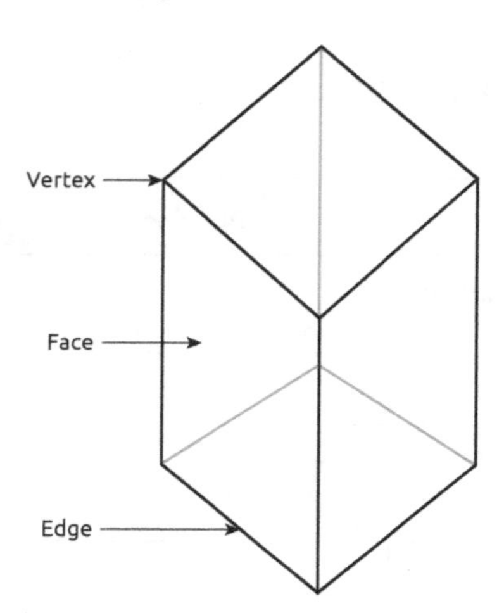

In the diagram, identify each part indicated and explain your reasoning.

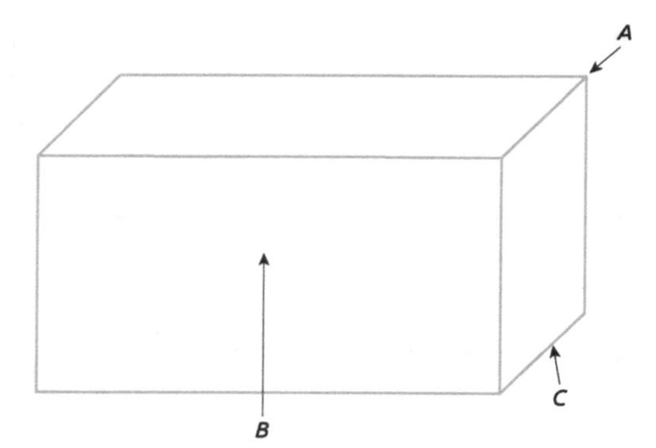

ANSWER: A is pointing at the point where three faces intersect, which is called a vertex. B is pointing at a flat edge of the solid, which is called a face. C is pointing at the segment where two faces intersect, which is called an edge.

Complete the activities below to check your understanding of the lesson content.

1. B is halfway between A and C in the diagram. What is the length of AC?

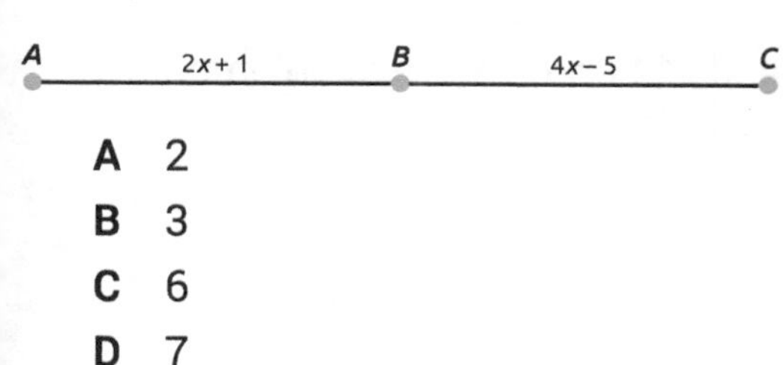

   A 2
   B 3
   C 6
   D 7
   E 14

2. If $m\angle N = 91°$, what type of angle is $\angle N$?
   A acute
   B right
   C obtuse
   D straight
   E vertical

Questions 3 and 4 refer to the information below.

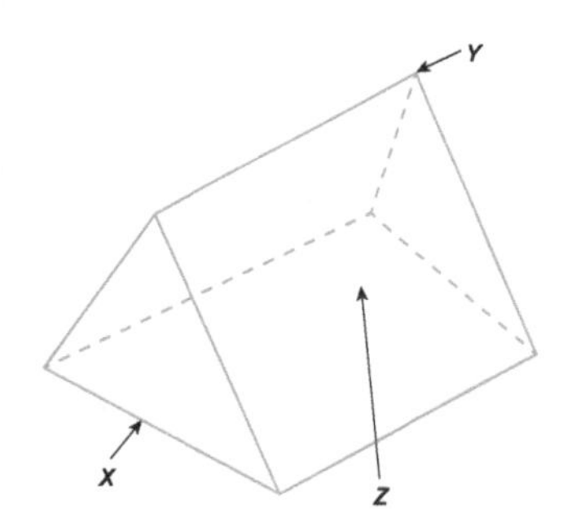

3. What is the above object called? Why?
   A a solid, because it is 2-dimensional
   B a solid, because it is 3-dimensional
   C a figure, because it is 2-dimensional
   D a rectangle, because it has 4 sides
   E a triangle, because it has 3 sides

4. What are X, Y, and Z called in the diagram?
   A X is a face, Y is an edge, and Z is a vertex.
   B X is an edge, Y is a face, and Z is a vertex.
   C X is a vertex, Y is an edge, and Z is a face.
   D X is an edge, Y is a vertex, and Z is a face.
   E X is a face, Y is a vertex, and Z is an edge.

Questions 5–7 refer to the information below.

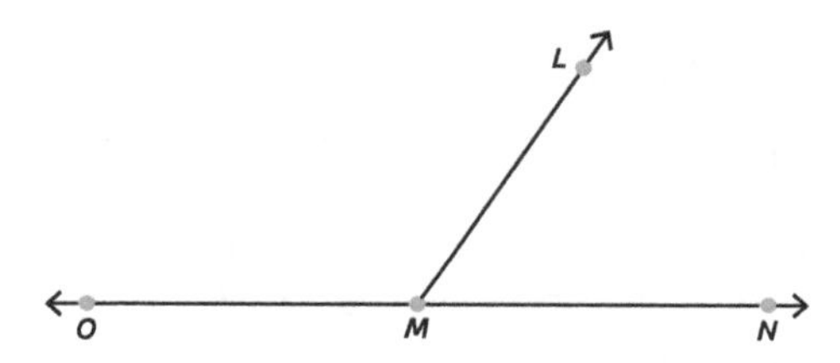

5. In the figure, what is LMN called?
   A line
   B line segment
   C ray
   D angle
   E point

6. In the figure, what is $\overrightarrow{MN}$ called?
   A line
   B line segment
   C ray
   D angle
   E point

7. In the figure, what is $\overleftrightarrow{ON}$ called?
   A line
   B line segment
   C ray
   D angle
   E point

See page 128 for answers and help.

TEST STRATEGY

Cross out wrong answer choices as you identify them. This will help to narrow answers down if you are unsure of the correct choice.

KEY POINT!

In geometry, knowing the terms and understanding what they mean is very important.

Solving Angle Measure, Perimeter, Area, Surface Area, and Volume Problems

SKILLS TIP

Adjacent angles and vertical angles are formed when two lines cross.

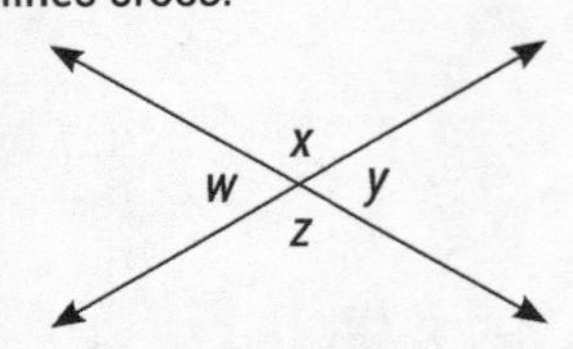

Vertical angles are equal: $\angle w$ and $\angle y$ are vertical angles.

Adjacent angles share a side: $\angle x$ and $\angle y$ are adjacent angles.

In this case, $\angle x$ and $\angle y$ are also supplementary angles.

Angle Measurement

Two angles whose sum is 90° are **complementary angles**. Two angles whose sum is 180° are **supplementary angles**.

If $m\angle A = (3x + 15)°$, and $m\angle B = (2x + 15)°$, and $\angle A$ and $\angle B$ are supplementary angles, what are the measurements of each angle?

ANSWER:

$$\begin{aligned} m\angle A + m\angle B &= 180° \\ (3x+15)+(2x+15) &= 180 \\ 5x+30 &= 180 \\ 5x &= 150 \\ x &= 30 \end{aligned}$$

Use $x = 30$ to find the measures of angles A and B.

$m\angle A = 3x + 15 = 3(30) + 15 = 105°$

$m\angle B = 2x + 15 = 2(30) + 15 = 75°$

Triangles

A triangle is a three-sided shape whose angles always add up to 180°.

In ΔABC, $m\angle A = (3x + 5)°$, and $m\angle B = (6x - 20)°$, and $m\angle C = (4x)°$. What are the measurements of each angle?

ANSWER:

$$\begin{aligned} m\angle A + m\angle B + m\angle C &= 180° \\ (3x+5)+(6x-20)+4x &= 180 \\ 13x-15 &= 180 \\ 13x &= 195 \\ x &= 15 \end{aligned}$$

Use $x = 15$ to find the measures of the angles.

$m\angle A = 3x + 5 = 3(15) + 5 = 50°$

$m\angle B = 6x - 20 = 6(15) - 20 = 70°$

$m\angle C = 4x = 4(15) = 60°$

Perimeter and Area

The perimeter of a figure is the distance around the figure. It can usually be determined by adding the length of each side of the figure. Consider the table below to review the formulas for finding the perimeter of some common shapes. Note that the perimeter of a circle is called the circumference.

Rectangle	Square	Circle
l = length; w = width	s = length	r = radius
$P = 2l + 2w$	$P = 4s$	$C = 2\pi r$ or $C = \pi d$

Find the circumference of a circle whose diameter is 12 inches. Write the answer in terms of π.

ANSWER: The formula for circumference is $C = \pi d$. Substitute to find the circumference, $C = \pi(12)$. The circumference is 12π inches.

The area of a figure is the number of square units it takes to cover a figure. Different area formulas are used for various types of figures. The following table shows the formulas for finding the area of some common shapes.

Rectangle	Square	Triangle	Circle
l = length, w = width	s = length	h = height, b = base	r = radius
$A = l \cdot w$	$A = s^2$	$A = \frac{1}{2}\ bh$	$A = \pi r^2$

Find the area of a triangle whose base is 9 feet and height is 6 feet.

ANSWER: Use the formula for area, $A = \frac{1}{2}\ bh$.

$$A = \frac{1}{2}\ bh$$
$$= \frac{1}{2}\ (9)(6)$$
$$= 27 \text{ ft}^2$$

What is the area of the shaded region in the following diagram?

ANSWER: Subtract the smaller rectangle from the larger rectangle.

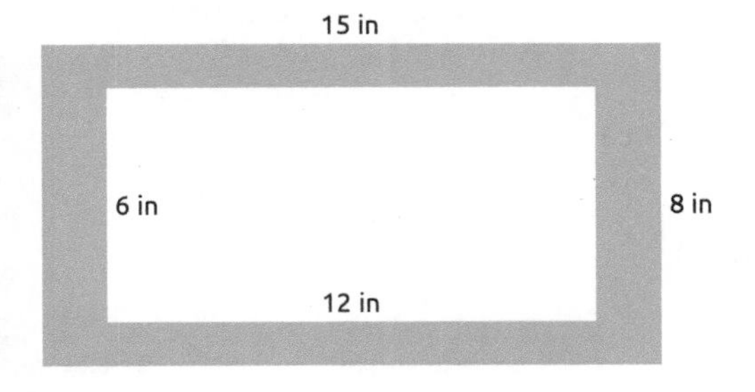

$$A = 15 \cdot 8 - 12 \cdot 6$$
$$= 120 - 72$$
$$= 48 \text{ in}^2$$

SKILLS TIP

You can use area to understand density. For example, Austin, TX, and Charlotte, NC, are both about 298 square miles. If Austin has about 790,000 people and Charlotte has about 731,000 people, which is most dense? Population density is defined by people per square mile, so Austin is more dense.

Solving Angle Measure, Perimeter, Area, Surface Area, and Volume Problems

SKILLS TIP

The perimeter of a figure is measured in units. The area of a figure is measured in square units. The volume of an object is measured in cubic units.

Volume

In a 3-dimensional object, the **volume** of the object is the number of cubic units it takes to fill the inside of the object. The volume of many objects can be found by multiplying the area of the base times the height of the object.

Rectangular Prism	Cylinder	Pyramid or Cone	Sphere
height, width, length	r, h	h = height	r = radius
$V = l \cdot w \cdot h$	$V = \pi r^2 h$	$V = \frac{1}{3}(\text{area of base}) \cdot h$	$V = \frac{4}{3}\pi r^3$

Find the volume of a cylinder that has a radius of 10 meters and a height of 15 meters in terms of π.

ANSWER: Use the formula: $V = \pi r^2 h$

$$= \pi(10)^2(15)$$
$$= 1{,}500 \text{ m}^3$$

Surface Area

In a 3-dimensional object, the **surface area** of the object is the sum of all of the areas of its sides. The surface area of many objects can be found by adding the area of their bases to their lateral area. The **lateral area** can be found by multiplying the perimeter of the base by the height of the object.

Find the surface area of a cylinder that has a radius of 10 yards and a height of 15 yards in terms of π.

ANSWER: Add the area of the two bases to the lateral area.

$$SA = 2\pi r^2 + 2\pi rh$$
$$= 2\pi(10)^2 + 2\pi(10)(15)$$
$$= 200\pi + 300\pi$$
$$= 500\pi \text{ yd}^2$$

Complete the activities below to check your understanding of the lesson content.

Questions 1 and 2 refer to the information below.

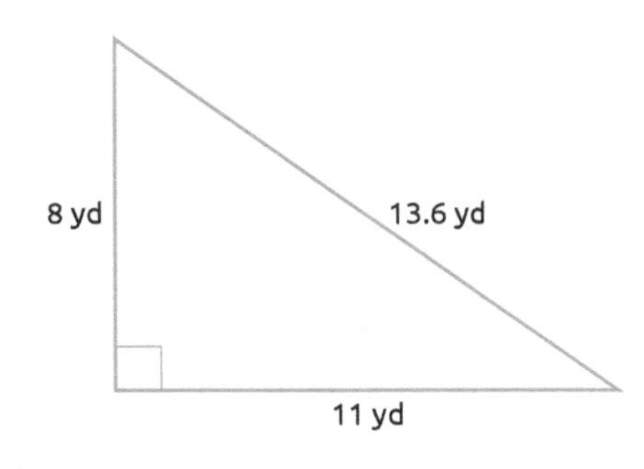

1. What is the perimeter of the triangle shown?
   - **A** 19 yards
   - **B** 24.6 yards
   - **C** 32.6 yards
   - **D** 44 yards
   - **E** 88 yards

2. What is the area of the triangle shown?
   - **A** 19 square yards
   - **B** 24.6 square yards
   - **C** 32.6 square yards
   - **D** 44 square yards
   - **E** 88 square yards

3. The area of a parallelogram can be found using the equation $A = bh$, where b is the base and h is the height. What is the area of the parallelogram shown?

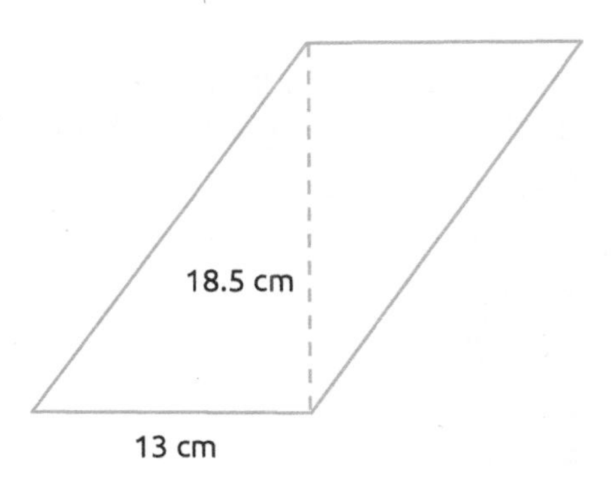

   - **A** 120.25 cm²
   - **B** 169 cm²
   - **C** 240.5 cm²
   - **D** 283 cm²
   - **E** 341.25 cm²

Questions 4 and 5 refer to the information below.

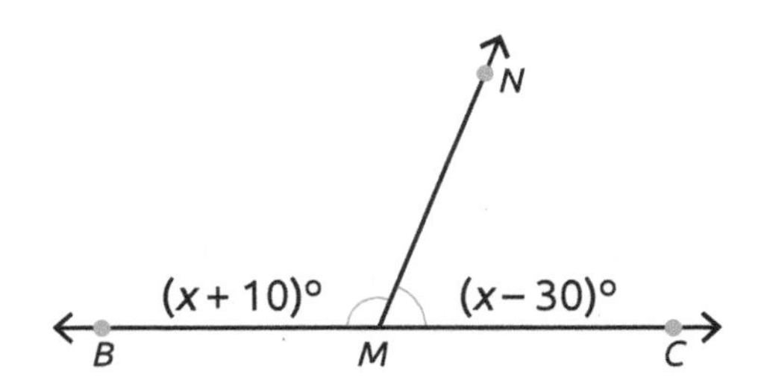

4. What are $\angle BMN$ and $\angle CMN$?
   - **A** acute angles
   - **B** right angles
   - **C** obtuse angles
   - **D** complementary angles
   - **E** supplementary angles

5. What is the measure of $\angle CMN$?
   - **A** 50°
   - **B** 70°
   - **C** 100°
   - **D** 110°
   - **E** 130°

KEY POINT!

A straight angle measures 180°. Supplementary angles are two angles whose sum is 180°. So, supplementary angles form a straight line.

TEST STRATEGY

Check to make sure the answer you choose is reasonable. Some answer choices may be eliminated because they do not make sense in the given situation.

KEY POINT!

The sum of the measurements of a triangle is always 180°.

Lesson Practice

Questions 6 and 7 refer to the information below.

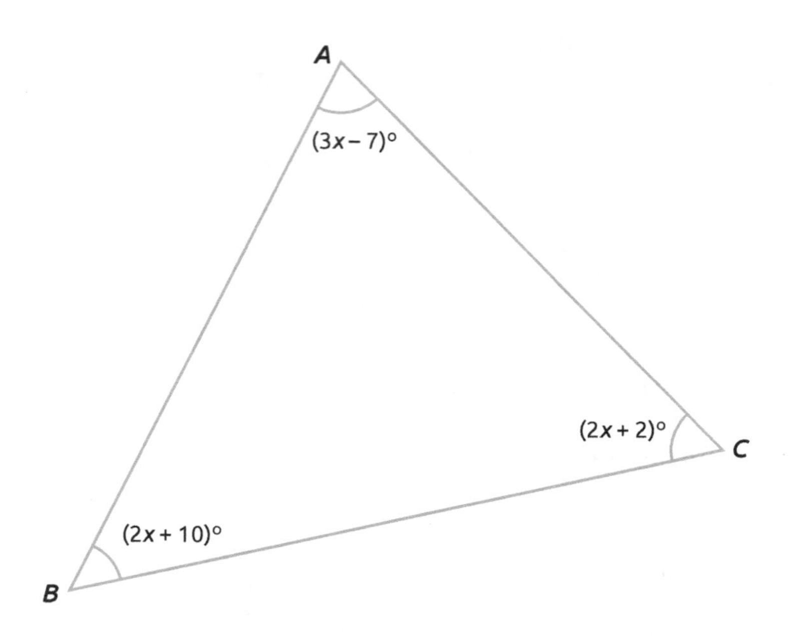

6. In ΔABC, what is the value of x?
   - **A** 10
   - **B** 25
   - **C** 52
   - **D** 60
   - **E** 68

7. In ΔABC, what is $m\angle A$?
   - **A** 10°
   - **B** 25°
   - **C** 52°
   - **D** 60°
   - **E** 68°

8. What is the measurement of the angle that is complementary to $\angle M$?

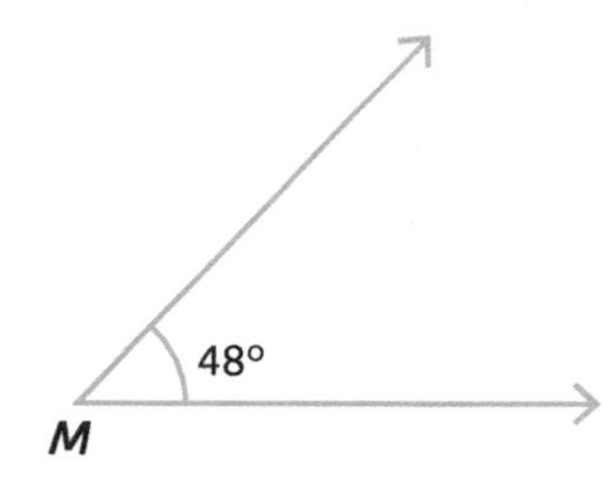

   - **A** 42°
   - **B** 48°
   - **C** 90°
   - **D** 132°
   - **E** 180°

9. $\angle X$ and $\angle Y$ are supplementary angles. If $m\angle X = (10n - 2)°$ and $m\angle Y = (2n + 2)°$, what is $m\angle X$?
   - **A** 15°
   - **B** 32°
   - **C** 90°
   - **D** 148°
   - **E** 180°

10. Which expression can be used to determine the area, in square feet, of the shaded region in the figure shown?

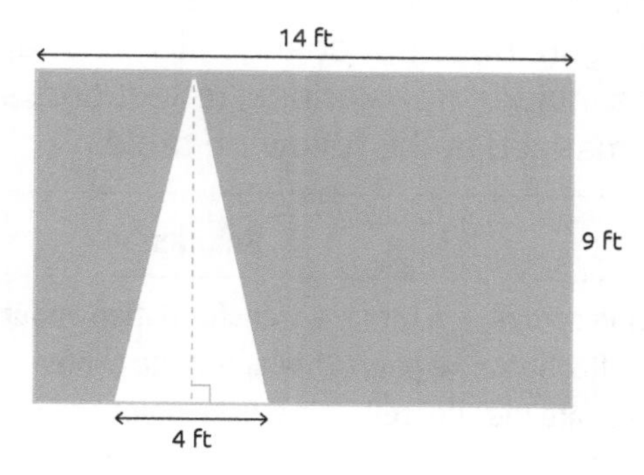

A $A = (14 \cdot 9) - (4 \cdot 9)$

B $A = (14 \cdot 9) - (\frac{1}{2} \cdot 4 \cdot 9)$

C $A = (14 \cdot 9) + (\frac{1}{2} \cdot 4 \cdot 9)$

D $A = (\frac{1}{2} \cdot 4 \cdot 9) - (14 \cdot 9)$

E $A = (\frac{1}{2} \cdot 4 \cdot 9) \cdot (14 \cdot 9)$

Questions 11 and 12 refer to the information below.

Isak is shipping the container shown.

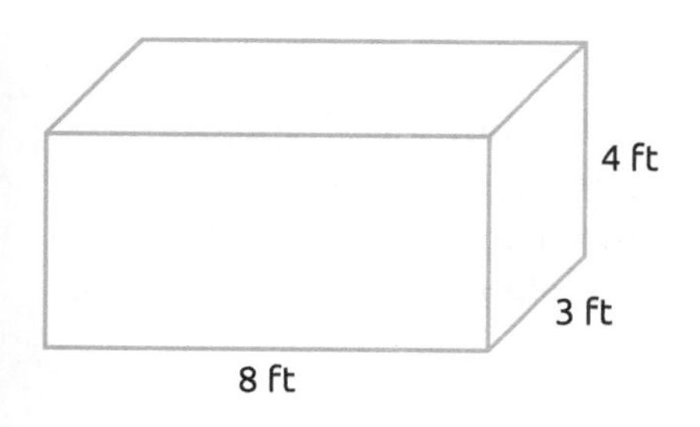

11. In the container, Isak will pack as many smaller boxes as he can. The smaller boxes measure 3 feet by 2 feet by 2 feet. How many of the smaller boxes can he fit in the larger container?

A 4

B 8

C 12

D 96

E 108

12. Isak wants to wrap the large container with shipping paper to protect it and needs to know how much to buy. What is the surface area of the container?

A 48 ft^2

B 96 ft^2

C 112 ft^2

D 136 ft^2

E 4,224 ft^2

13. A carnival game involves throwing darts at a circular target that has a diameter of 6 inches. What is the area of the target? Use 3.14 for π.

A 18.84 in^2

B 28.26 in^2

C 37.68 in^2

D 113.04 in^2

E 452.16 in^2

See page 128 for answers and help.

Understanding Similarity and Congruence

SKILLS TIP

Similar figures have the same shape but may be different sizes. For example, two similar triangles will have the same angle measurements, but the sides may be different lengths. In contrast, congruent figures have both the same shape and the same size.

Congruence

A **transformation** of a figure is a change in the position, size, or shape of the figure. The **preimage** is the figure before the transformation, and the **image** is the figure after the transformation. Some transformations are called congruence transformations because the figure does not change size or shape, only position. Congruence transformations include translations, reflections, and rotations. These transformations are summarized in the following table.

Translation	Reflection	Rotation
A slide – all points in a figure move the same direction and same distance.	A flip – a transformation across a line, called a line of reflection. Each point and its image are the same distance from the line of reflection.	A turn – a transformation about a point, *P*, which is the center of reflection. Each point and its image are the same distance from *P*.
Q P R Q' S P' R' S'	B B' A A' C C' D D'	L' K' K J' M' J L P M

Paco held up a message on a poster board that can be read by holding it up to a mirror. What transformation must occur in order to read the message? Explain your reasoning.

ANSWER: It is a reflection. The line of reflection is the mirror, and all points of the preimage and image are equal distances from the line of reflection, but the image is flipped over that line.

Similarity Transformation

A **dilation** is a transformation that changes the size but not the shape of the figure. The image and preimage are similar. The corresponding angles are still congruent, and the corresponding sides are proportional to each other. A dilation can be an enlargement or a reduction. The **scale factor** describes how much the preimage is enlarged or reduced.

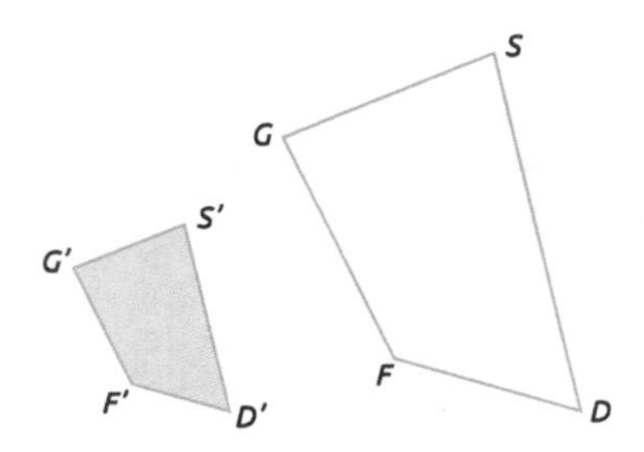

A scale factor is multiplied by each dimension of a preimage to find the dimensions of the dilated image. If a scale factor is greater than 1, then the dilation is an enlargement. If a scale factor is greater than 0 but less than 1, then the dilation is a reduction.

Rodney has a 4-inch by 6-inch photograph that he wants to enlarge to 6-inches by 9-inches. What is the scale factor he should use for the enlargement?

$$4 \times sf = 6$$
$$sf = \frac{3}{2}$$
$$6 \times sf = 9$$
$$sf = \frac{3}{2}$$

ANSWER: The scale factor is $\frac{3}{2}$ or 1.5.

The community park planning committee is drawing up plans for a new park. The following drawing lays out the plan for the park. If the actual dimensions of the shuffleboard court are 9 meters by 22.5 meters, what is the scale factor being used to create the drawing? What are the actual dimensions of the park?

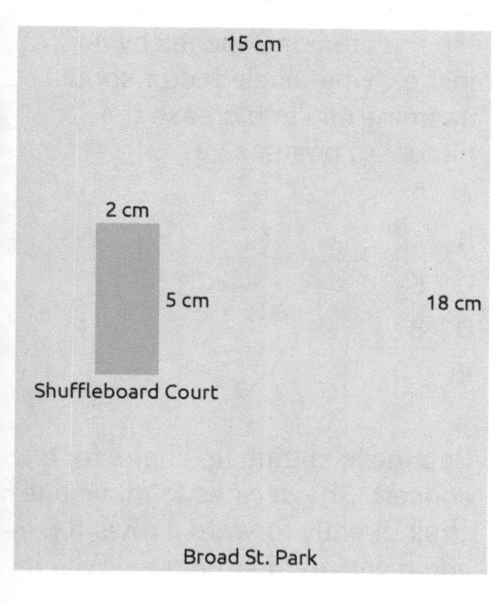

9 meters = 900 centimeters

$$900 \times sf = 2$$
$$sf = \frac{2}{900} = \frac{1}{450}$$

22.5 meters = 2250 centimeters

$$2250 \times sf = 5$$
$$sf = \frac{5}{2250} = \frac{1}{450}$$

The scale factor is $\frac{1}{450}$.

So, to find the actual dimensions of the park, multiply by the scale factor of $\frac{450}{1}$.

$\frac{450}{1}$ x 18= 8100 cm, or 81 m

$\frac{450}{1}$ x 15= 6750 cm, or 67.5 m

ANSWER: The dimensions of the park are 67.5 meters by 81 meters.

SKILLS TIP

If coordinates are given that describe the vertices of a preimage, then the vertices of the dilated image can be found by multiplying each coordinate by the scale factor. For example, the table below shows the coordinates for the preimage points A and B. To determine A' and B', each coordinate was multiplied by the scale factor 2.

Pre-Image	Image
A (1, 2)	A'(2, 4)
B (−3, −5)	B'(−6, −10)

Lesson Practice

KEY POINT!

When determining the type of transformation, look at the orientation of the labeled vertices of the preimage and the image.

Complete the activities below to check your understanding of the lesson content.

Questions 1–4 refer to the information below.

The following diagram shows several

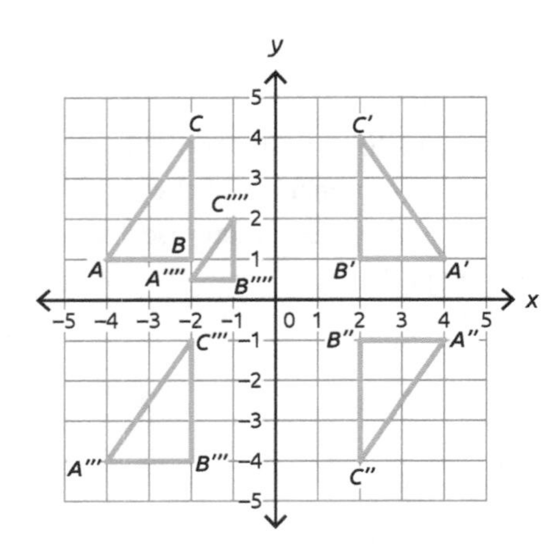

transformations of ΔABC.

1. Which triangle is not a congruence transformation of ΔABC?
   - **A** ΔABC
   - **B** ΔA′B′C′
   - **C** ΔA″B″C″
   - **D** ΔA‴B‴C‴
   - **E** ΔA⁗B⁗C⁗

2. Which triangle is a rotation of ΔABC?
   - **A** ΔBAC
   - **B** ΔA′B′C′
   - **C** ΔA″B″C″
   - **D** ΔA‴B‴C‴
   - **E** ΔA⁗B⁗C⁗

3. Which triangle is translation of ΔABC?
   - **A** ΔABC
   - **B** ΔA′B′C′
   - **C** ΔA″B″C″
   - **D** ΔA‴B‴C‴
   - **E** ΔA⁗B⁗C⁗

4. Which triangle is a reflection of ΔABC?
   - **A** ΔABC
   - **B** ΔA′B′C′
   - **C** ΔA″B″C″
   - **D** ΔA‴B‴C‴
   - **E** ΔA⁗B⁗C⁗

5. Jasmina has a picture that she wants to enlarge to a poster that measures 24 inches by 32 inches. If her picture is 3 inches by 4 inches, what scale factor should Jasmine use to increase the picture to poster size?
   - **A** 2
   - **B** 4
   - **C** 5
   - **D** 8
   - **E** 10

6. Daphne is setting up chairs for a concert. She decides to move one chair directly forward 3 rows to add a seat in the front row. Which transformation did Daphne just perform with the chair?
   - **A** translation
   - **B** reflection
   - **C** rotation
   - **D** dilation
   - **E** reproduction

Questions 7 and 8 refer to the information below.

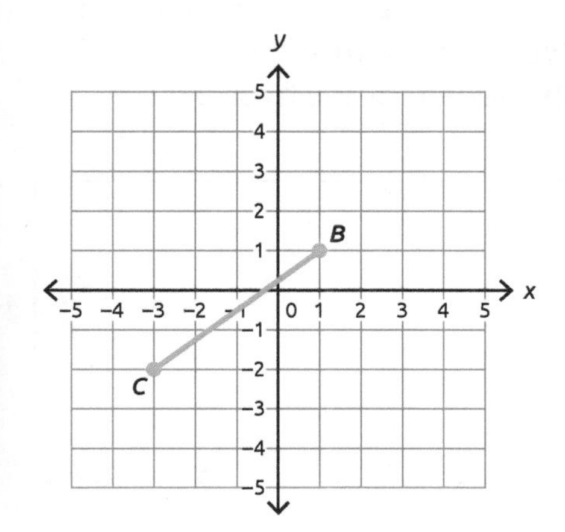

7. If $\overline{BC}$ is dilated by a scale factor of 2, what is the coordinate of point B'?

   A $\left(\frac{1}{2}, \frac{1}{2}\right)$

   B (2, 2)

   C (−2, −2)

   D (−6, −4)

   E $\left(-\frac{3}{2}, -1\right)$

8. If the image of $\overline{B'C'}$ has coordinates of $B'(3, -2)$ and $C'(-1, -5)$, which of the following transformations could have occurred?

   A a dilation with scale factor 2

   B a rotation about the origin

   C a translation down 3 units and right 2 units

   D a reflection across the *y*-axis

   E a reflection across the *x*-axis

See page 128 for answers and help.

TEST STRATEGY

Read the problem carefully. Sometimes, it can be helpful to draw a diagram of what is happening if no diagram was provided.

Using the Pythagorean Theorem

SKILLS TIP

Remember, simplify square roots by separating out any perfect squares and taking the square root to simplify. For example,

$\sqrt{12} = \sqrt{4 \cdot 3} = \sqrt{4} \cdot \sqrt{3} = 2\sqrt{3}$

Pythagorean Theorem

A right triangle has two legs and a **hypotenuse**. The hypotenuse is the longest side, directly across from the right angle. The **Pythagorean Theorem** says that the sum of the squares of each leg is equal to the square of the hypotenuse. This is illustrated in the following table.

Pythagorean Theorem:	
$a^2 + b^2 = c^2$ where a and b are the legs and c is the hypotenuse of a right triangle	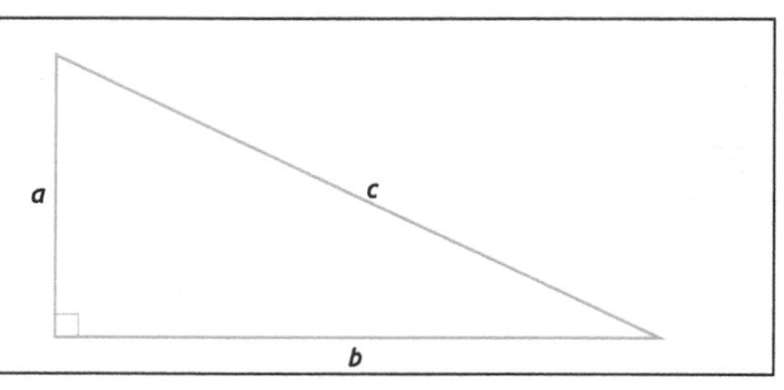

In the triangle shown here, what is the value of x?

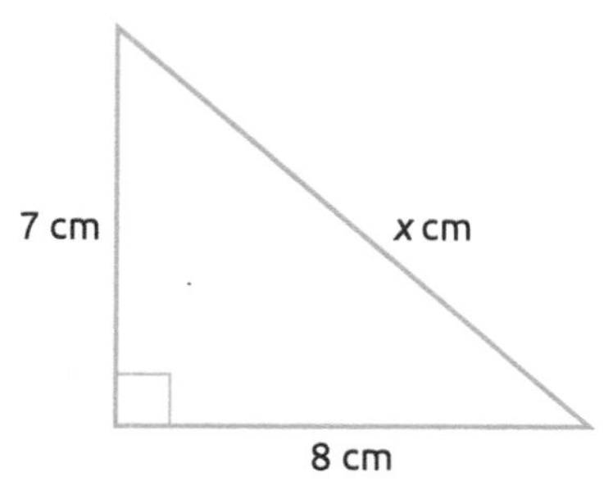

$$a^2 + b^2 = c^2$$
$$7^2 + 8^2 = x^2$$
$$113 = x^2$$
$$\sqrt{113} = x$$

ANSWER: The value of x is $\sqrt{113}$ centimeters.

The relationship represented by the Pythagorean Theorem is true for all right triangles.

Candace and Jerome are surveying a road that will be built on a hill. They are working to determine the total drop from the top of the hill to the bottom. The following diagram represents what they know. How far is the drop in the hill?

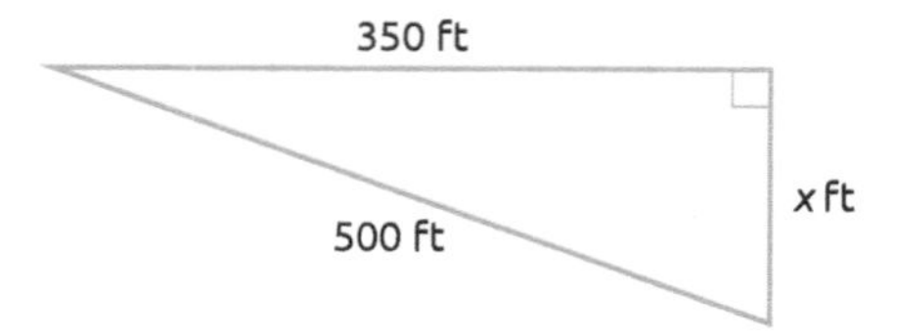

$$a^2 + b^2 = c^2$$
$$350^2 + x^2 = 500^2$$
$$122500 + x^2 = 250000$$
$$x^2 = 127500$$
$$x = \sqrt{127500}$$
$$x = \sqrt{2500 \times 51}$$
$$x = 50\sqrt{51} \text{ ft}$$

ANSWER: The drop is $50\sqrt{51}$ feet, or about 357 feet.

Complete the activities below to check your understanding of the lesson content.

1. Hector is painting the border of a picture on a wall. He wants to make sure that the corner at angle *A* is a right angle. What value must *x* be to ensure this?

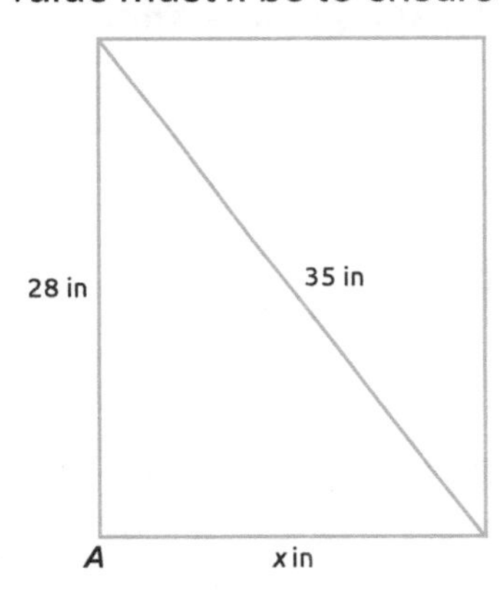

A 7
B 21
C 45
D 63
E 84

2. The drawing shows where three friends are standing in relation to one another. How far away is Jessa from Yong, to the nearest tenth of a meter?

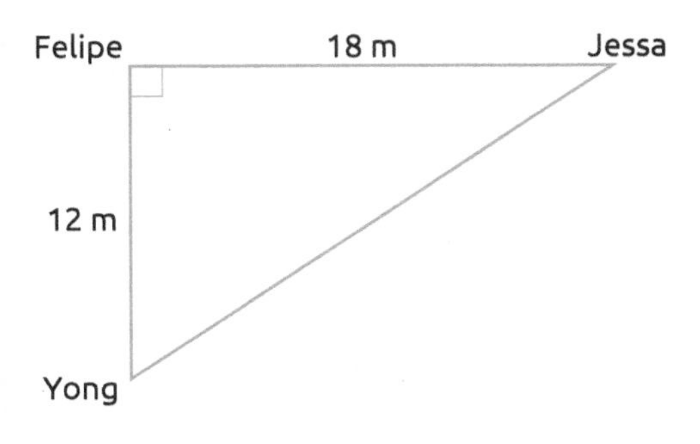

A 6.0 meters
B 13.4 meters
C 21.6 meters
D 180.0 meters
E 468.0 meters

3. The sail on a boat is a right triangle with the measurements shown in the drawing. Which of the following expressions could be used to find *a*, the length of the sail that attaches to the mast?

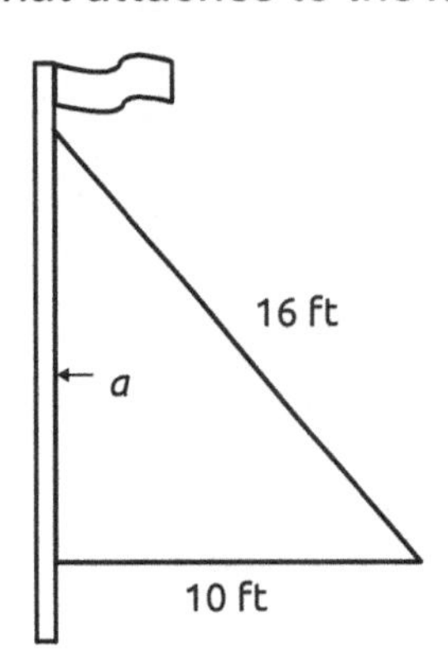

A $16^2 - 10^2$
B $\sqrt{16 - 10}$
C $\sqrt{16^2 - 10^2}$
D $(16 - 10)^2$
E $(16 + 10)^2$

See page 128 for answers and help.

KEY POINT!

When working with right triangles, the Pythagorean Theorem and trigonometric ratios can all be used to determine missing side lengths.

Answer the questions based on the content covered in this unit.

1. N is halfway between M and O in the following diagram. What is the length of MO?

M $2x + 5$ N $4x - 3$ O

 A 2 units
 B 4 units
 C 13 units
 D 26 units
 E 28 units

2. If $m\angle Z = 90°$, what type of angle is $\angle Z$?

 A acute
 B right
 C obtuse
 D straight
 E supplementary

Questions 3–6 refer to the transformations of rectangle ABCD shown below.

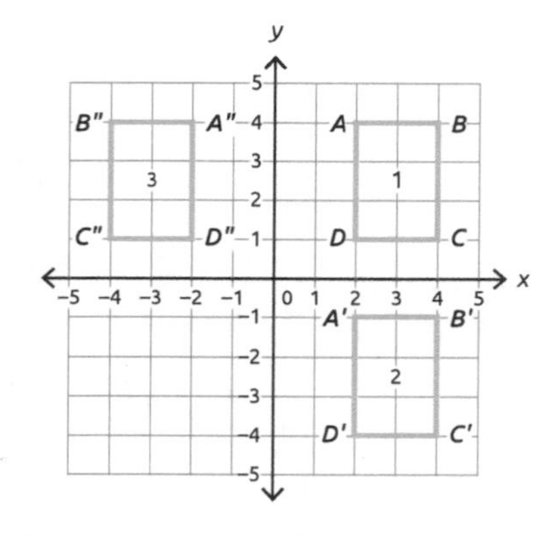

3. Which lists all the congruence transformations of rectangle *ABCD*?

 A only rectangle 1
 B only rectangle 2
 C only rectangle 3
 D rectangles 2 and 3
 E No congruence transformations are shown.

4. Which is a rotation of rectangle *ABCD*?

 A rectangle 1
 B rectangle 2
 C rectangle 3
 D rectangles 2 and 3
 E none

5. Which is a translation of rectangle *ABCD*?

 A rectangle 1
 B rectangle 2
 C rectangle 3
 D rectangles 2 and 3
 E none

6. Which is a reflection of rectangle *ABCD*?

 A rectangle 1
 B rectangle 2
 C rectangle 3
 D rectangles 2 and 3
 E none

Questions 7 and 8 refer to the information below.

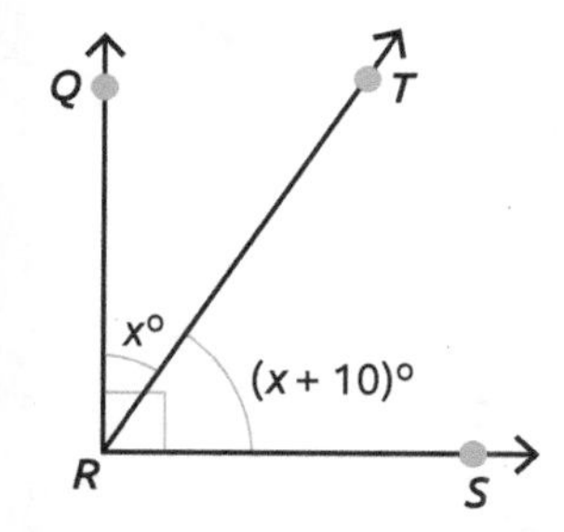

7. What are $\angle QRT$ and $\angle SRT$?
   - **A** straight angles
   - **B** right angles
   - **C** obtuse angles
   - **D** complementary angles
   - **E** supplementary angles

8. What is the measure of $\angle SRT$?
   - **A** 30°
   - **B** 40°
   - **C** 50°
   - **D** 60°
   - **E** 90°

9. Makiah is playing a game with cards. The game begins when she takes the top card, which is face down, and flips it over onto the discard pile. Which transformation does Makiah perform to begin the game?

Discard

   - **A** translation
   - **B** reflection
   - **C** rotation
   - **D** dilation
   - **E** reproduction

Questions 10 and 11 refer to the information below.

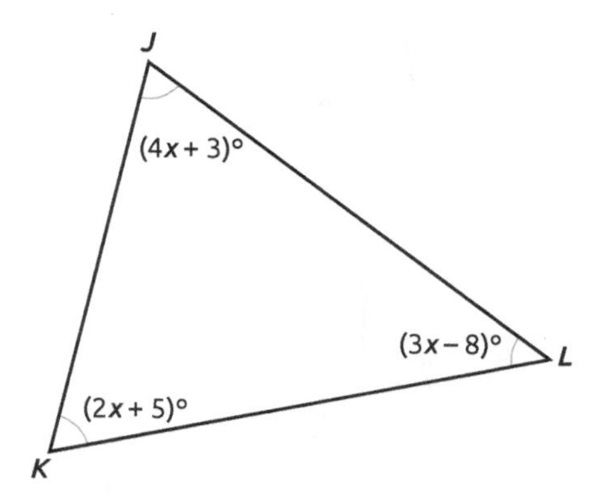

10. In ΔJKL, what is the value of x?
    - **A** 20
    - **B** 45
    - **C** 52
    - **D** 83
    - **E** 180

11. In ΔJKL, what is $m\angle L$?
    - **A** 20°
    - **B** 45°
    - **C** 52°
    - **D** 83°
    - **E** 180°

12. What is the measure of the angle that is complementary to $\angle E$ shown below?

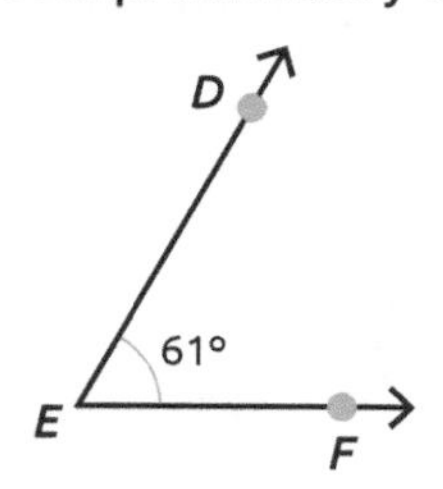

    - **A** 29°
    - **B** 61°
    - **C** 90°
    - **D** 119°
    - **E** 180°

Questions 13 and 14 refer to the information below.

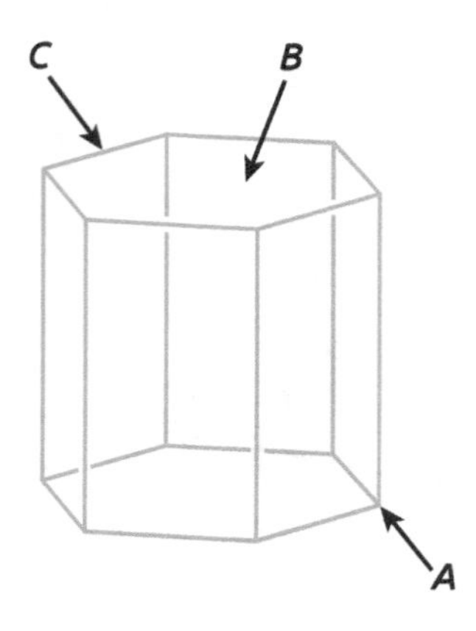

13. What is the object called? Why?
    - **A** a solid, because it is 2-dimensional
    - **B** a solid, because it is 3-dimensional
    - **C** a figure, because it is 2-dimensional
    - **D** a rectangle, because it has 4 sides
    - **E** a triangle, because it has 3 sides

14. What are *A*, *B*, and *C* called in the diagram?
    - **A** *A* is a vertex, *B* is a face, and *C* is an edge.
    - **B** *A* is an edge, *B* is a face, and *C* is a vertex.
    - **C** *A* is a vertex, *B* is an edge, and *C* is a face.
    - **D** *A* is an edge, *B* is a vertex, and *C* is a face.
    - **E** *A* is a face, *B* is a vertex, and *C* is an edge.

Questions 15 and 16 refer to the information below.

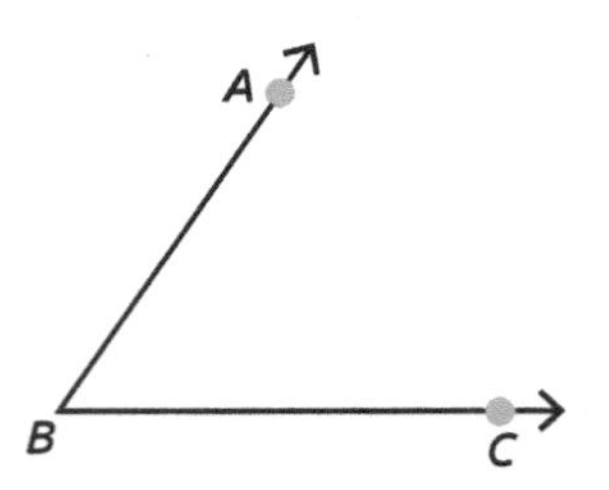

15. In the figure, what is *ABC* called?
    - **A** point
    - **B** angle
    - **C** ray
    - **D** line
    - **E** line segment

16. In the figure, what is $\overrightarrow{BA}$ called?
    - **A** point
    - **B** angle
    - **C** ray
    - **D** line
    - **E** line segment

17. Jacob is painting a mural on the side of a community center. He wants to make sure that the corner at angle *A* is a right angle. What value must *x* be to ensure this?

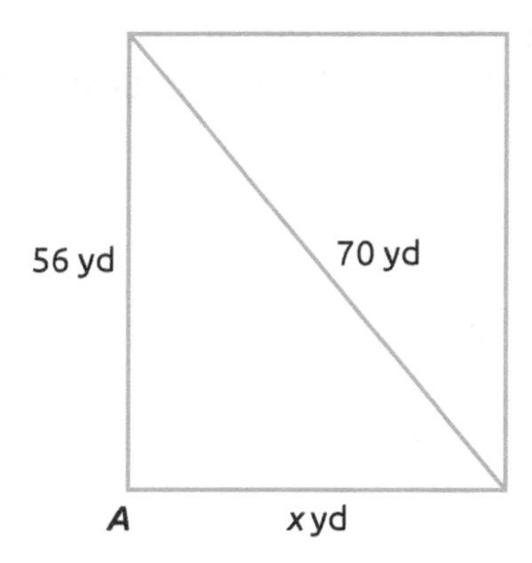

- **A** 14 yards
- **B** 36 yards
- **C** 42 yards
- **D** 90 yards
- **E** 126 yards

Questions 18 and 19 refer to the information below.

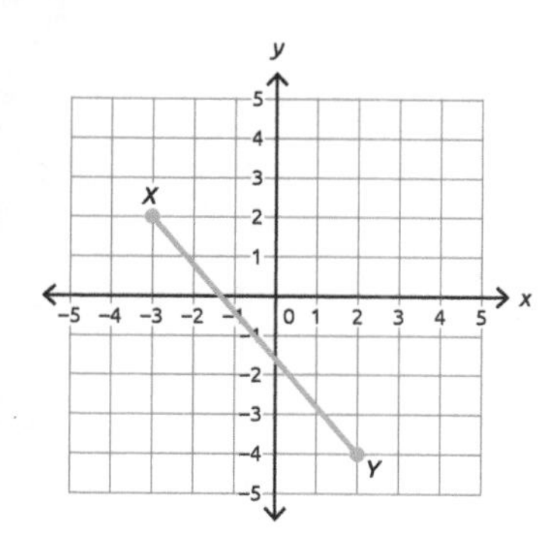

18. If $\overline{XY}$ is dilated by a scale factor of $\frac{1}{2}$, what is the coordinate of the image X'?

  A $(4, -8)$
  B $(1, -2)$
  C $(-6, 4)$
  D $\left(-\frac{3}{2}, -1\right)$
  E $\left(-\frac{3}{2}, 1\right)$

19. If the image of $\overline{X'Y'}$ has coordinates of $X'(-3, -2)$ and $Y'(2, 4)$, which of the following transformations could have occurred?

  A a dilation with scale factor 2
  B a rotation about the origin
  C a translation down 6 units and right 2 units
  D a reflection across the x-axis
  E a reflection across the y-axis

Questions 20 and 21 refer to can of vegetable juice shown below.

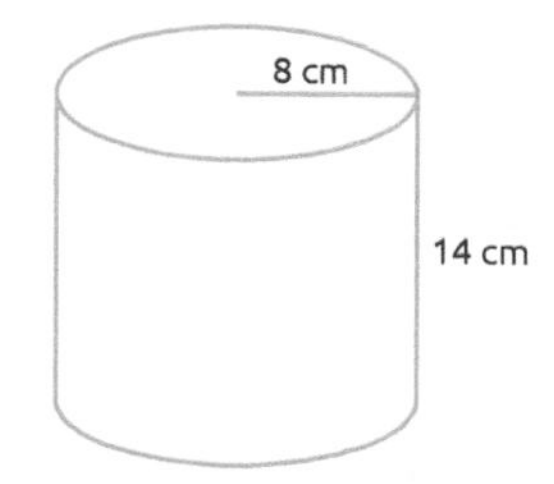

20. Charisse drank one-fourth of the juice in the container. How many cubic centimeters of juice does she have left? Use 3.14 for π.

  A 351.68 cm³
  B 703.36 cm³
  C 1,406.72 cm³
  D 2,110.08 cm³
  E 2,813.44 cm³

21. A label covers the lateral surface of Charisse's can. Which of the following expressions can be used to determine the area of the label?

  A $A = 2\pi rh$
  B $A = \pi r^2 h$
  C $A = 2\pi r^2 + 2\pi rh$
  D $A = \pi rh$
  E $A = 2\pi r^2 + \pi rh$

22. Jermaine has a photograph that he wants to enlarge to a size of 8 inches by 10 inches. If his original is 4 inches by 6 inches, how many times larger does the picture need to be to get closest to the enlarged size?

  A $\frac{1}{2}$
  B 2
  C 4
  D 6
  E 8

23. The drawing shows where the library and school are in relation to Rachel's home. To the nearest tenth-mile, how far is it from her home to the school?

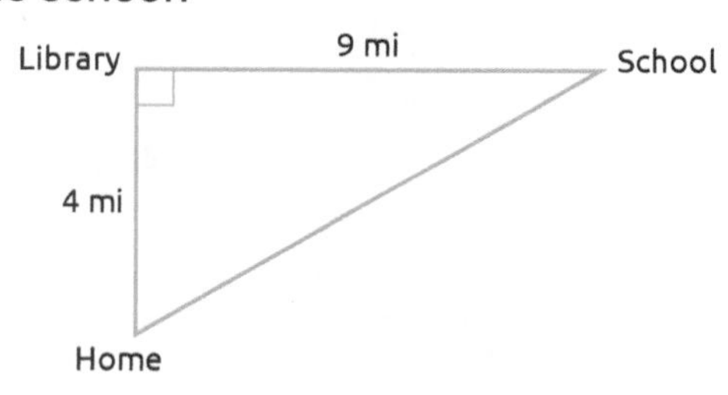

  A 5.0 miles
  B 6.2 miles
  C 8.1 miles
  D 9.8 miles
  E 13.0 miles

See page 128 for answers.

Lesson 1

1. **E.** Set both expressions equal to each other and solve for x. Then, substitute $x = 3$ into both expressions and add them together.

2. C.

3. B.

4. **D.** *X* is an edge because it is a segment where two faces intersect. *Y* is a vertex because it is a point where three faces intersect. *Z* is a face because it is a flat edge of the solid.

5. **D.** An angle is two rays that share a common endpoint. *M* is the common endpoint for this figure.

6. **C.** A ray has one endpoint and continues indefinitely in one direction.

7. **A.** A line is a straight path that extends indefinitely in both directions.

Lesson 2

1. **C.** The perimeter of the triangle is the sum of all 3 sides.

2. **D.** The area of a triangle is one-half times the base times the height. The base is 11 yards, and the height is 8 yards.

3. C.

4. **E.** The angles together form a straight angle, which measures 180°.

5. B.

6. **B.** The sum of the three angles is 180°. Set up the equation $(2x + 10) + (2x + 2) + (3x - 7) = 180$, and then solve for x.

7. **E.** Since $x = 25$, $m\angle A = 3(25) - 7 = 68°$.

8. **A.** Complementary angles add up to 90°. $90° - 48° = 42°$.

9. D.

10. **B.** Find the area of the rectangle and subtract from it the area of the triangle.

11. **B.** Divide the volume of the large container by the volume of a smaller box.

12. D.

13. **B.** Use the formula, $A = \pi r^2$, where the radius is 3 inches.

Lesson 3

1. **E.** $\Delta A''''B''''C''''$ is a reduction dilation, which is a similarity transformation.

2. **C.** $\Delta A''B''C''$ is a rotation of ΔABC around the origin.

3. **D.** $\Delta A'''B'''C'''$ is a translation of ΔABC. The original triangle is shifted 5 units down.

4. **B.** $\Delta A'B'C'$ is a reflection of ΔABC across the *y*-axis.

5. D.

6. **A.** The movement of the chair is like a translation. The chair changes position but not size or orientation.

7. **B.** Multiply the coordinates of the preimage by 2.

8. C.

Lesson 4

1. **B.** Use the Pythagorean Theorem to find x.

2. **C.** Use the Pythagorean Theorem to find the hypotenuse of the triangle.

3. **C.** The expression $\sqrt{16^2 - 10^2}$ uses the Pythagorean Theorem to find x.

Unit Test

1. D.
2. B.
3. D.
4. E.
5. B.
6. C.
7. D.
8. C.
9. B.
10. A.
11. C.
12. A.
13. B.
14. A.
15. B.
16. C.
17. C.
18. E.
19. D.
20. D.
21. A.
22. B.
23. D.

- **acute angle** – an angle whose measurement, *A*, is $0° \leq A < 90°$
- **angle** – two rays sharing a common endpoint
- **complementary angles** – two angles whose sum is 90°
- **dilation** – a transformation that changes the size but not the shape of the figure
- **edge** – the segment formed where two faces of a solid intersect
- **endpoint** – points at each end of a line segment or one end of a ray
- **face** – each flat surface of a solid
- **hypotenuse** – the longest leg of a right triangle directly across from the right angle
- **image** – the image of a figure after a transformation
- **lateral area** – the area of the side(s) of an object between the bases
- **line** – straight path made up of an infinite number of points; has no thickness and extends in both directions indefinitely
- **line segment** – a part of a line that consists of two endpoints and all of the points between them
- **obtuse angle** – an angle whose measurement, *A*, is $90° < A \leq 180°$
- **plane** – a flat surface with no thickness that extends forever in each direction
- **point** – names a location but has no size; represented by a dot
- **preimage** – the image of a figure before a transformation
- **Pythagorean Theorem** – a relationship in all right triangles that says the sum of the squares of each leg is equal to the square of the hypotenuse
- **ray** – a part of a line that has one endpoint and extends forever in one direction
- **reflection** – a flip or a transformation across a line, called a line of reflection; each point and its image are the same distance from the line of reflection
- **right angle** – an angle whose measurement, *A*, is $A = 90°$
- **rotation** – a turn; a transformation about a point, *P*, which is the center of reflection; each point and its image are the same distance from *P*
- **scale factor** – describes how much the preimage is enlarged or reduced
- **supplementary angles** – two angles whose sum is 180°
- **surface area** – the number of square units it takes to cover the outside surfaces of a 3-dimensional object
- **transformation** – a change in the position, size, or shape of a figure
- **translation** – a slide; when all points in a figure are moved the same direction and same distance
- **vertex** – a point that is formed by the intersection of three or more faces of a solid
- **volume** – the number of cubic units it takes to fill a 3-dimensional object

Study More!

Know Shapes and Their Attributes

- Know the names and attributes of various shapes with different numbers of sides.

Solve Angle Measure, Perimeter, Area, Surface Area, and Volume Problems

- Know how to find the sum of the measures of the interior angles of various shapes with different numbers of sides.
- Know how to find the perimeter and area of composite figures.
- Know how to find the surface area and volume of other objects.

Understand Similarity and Congruence

- Understand key similarity ratios of 30°-60°-90° and 45°-45°-90° triangles.
- Continue to perform more congruence and similarity transformations on a coordinate plane.

Use the Pythagorean Theorem

- Understand how to find missing length of a triangle's leg by using the Pythagorean Theorem.

Measurement and Data

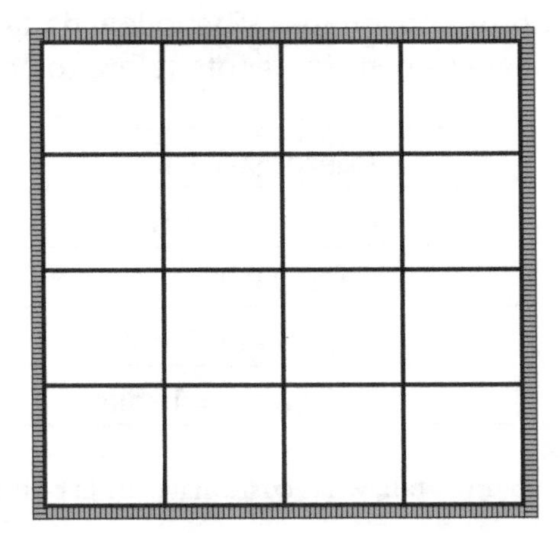

Measurement is a valuable tool and skill for life and work. Whether the task involves building something, painting a wall, or even hanging a picture, the ability to measure or estimate a measurement is useful. Consider the following scenario:

Fernando is designing a tile patio similar to what is shown above with large tiles surrounded by smaller tile border. Each large tile measures 36 inches on each side, and the width of the border is 2 inches. What is the perimeter of the patio, in inches?

Answer: 592 inches

KEY WORDS

- English units of measurement
- metric units of measurement
- rate
- unit multiplier

Measuring and Estimating Lengths in Standard Units

Customary Units of Length

The United States is one of the few countries that uses customary units to measure length. Length is based on units called the inch (in), foot (ft), yard (yd), and mile (mi), from shortest to longest. The following table demonstrates how the **English units of measurement** for length relate to each other.

Customary Units
12 inches = 1 foot
3 feet = 1 yard
5,280 feet = 1 mile
1,760 yards = 1 mile

If a man is said to be 6 units tall, what customary units of measurement are most likely being used?

ANSWER: The units being used here are feet. Inches are too small, and yards are too large.

Cedric measured a space that needs a 2-by-4 piece of lumber for support. The space measured 56 inches. He has 2-by-4 lumber in lengths of 4 feet, 6 feet, and 8 feet. Which length should he select to minimize waste? Explain.

ANSWER: 56 inches ÷ 12 inches = $4\frac{2}{3}$, or 4 feet and 8 inches. The 4-foot piece is too short. The 6-foot piece of lumber minimizes waste.

Metric Units of Length

Much of the world uses **metric units of measurement** for length, which include centimeter (cm), meter (m), and kilometer (km) as the primary units. Consider the following table to see how the units within the metric system compare to each other.

Metric Units
1,000 millimeter = 1 meter
100 centimeters = 1 meter
10 decimeters = 1 meter
10 meters = 1 decameter
100 meters = 1 hectometer
1,000 meters = 1 kilometer

If a room is said to be 10 units long, what metric units of measurement are most likely being used?

ANSWER: The units being used here are meters. Centimeters are too small, and kilometers are too large.

Titus is driving to school. It usually takes him about 10 minutes to get to school. Which metric unit is most likely to be used to describe the distance Titus lives from school? Explain.

ANSWER: Kilometers are the units that best describe longer distances such as the distance from his home to his school.

Measuring

Many different instruments can be used to measure lengths. Some of the instruments include a ruler, a yard/meter stick, a tape measure, or an odometer in a vehicle. It is important when using these tools to identify the units being used. Typically, the instrument will label what the units are that it is measuring. The ruler below measures centimeters along the top edge and inches along the bottom edge, as the label indicates.

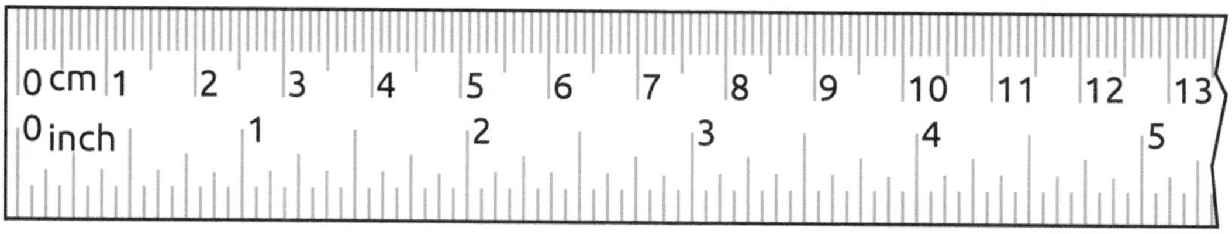

0 cm 1 2 3 4 5 6 7 8 9 10 11 12 13
0 inch 1 2 3 4 5

In the diagram below, what is the approximate length of the yarn shown, in inches and centimeters?

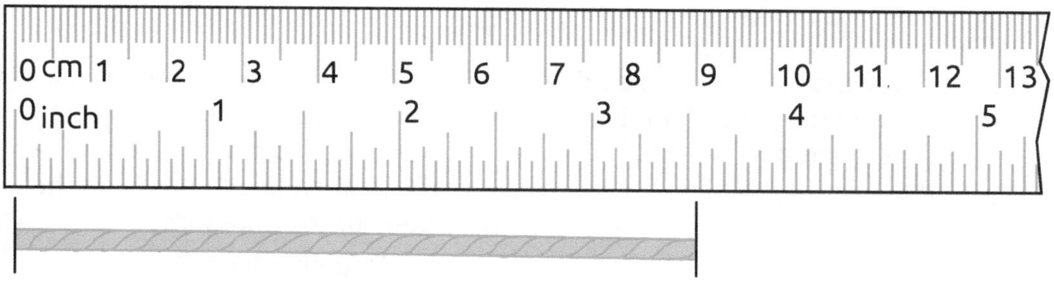

0 cm 1 2 3 4 5 6 7 8 9 10 11 12 13
0 inch 1 2 3 4 5

ANSWER: The yarn is about 9 cm, or $3\frac{1}{2}$ inches long.

SKILLS TIP

One advantage to metric units is that they differ from each other by one power of 10. When converting between metric units, simply multiply or divide by the correct power of 10. This can be done by moving the decimal point in the number.

Lesson Practice

TEST STRATEGY

Try to think of other things in life with which you are familiar in terms of length and distances. This may give you a reference point when selecting your answers.

KEY POINT!

When using a ruler to measure, be sure to read the units that the ruler measures.

Complete the activities below to check your understanding of the lesson content.

Questions 1–3 refer to the following diagram.

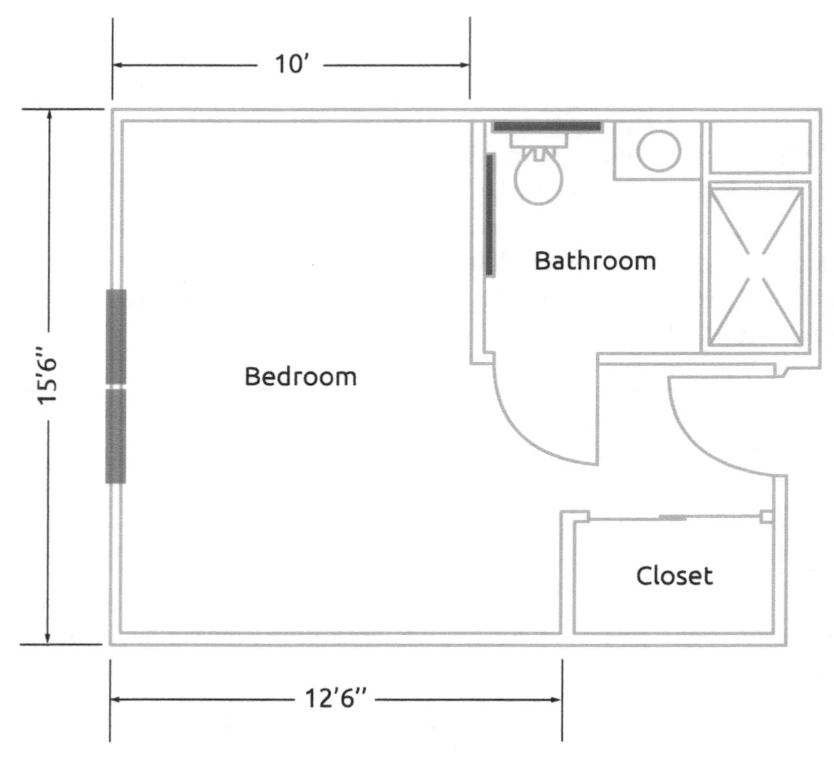

1. In the drawing, the closet is said to be 6 units long. What units are most likely being referred to?
   A centimeters
   B kilometers
   C yards
   D inches
   E feet

2. In the drawing, what is the most likely width of the sink in the bathroom?
   A 36 feet
   B 36 inches
   C 3 yards
   D 3 inches
   E 12 feet

3. Nick is planning to paint the walls of the bedroom. Which of the following instruments would be the best choice to measure the walls of the bedroom?
   A 12-inch ruler
   B yardstick
   C 25-foot tape measure
   D meterstick
   E 30-centimeter ruler

4. Xavier is looking at a rock wall he is preparing to climb at his indoor fitness club. Which of the following is a reasonable height for the rock wall?

   A 500 yards
   B 50 inches
   C 5 centimeters
   D 50 feet
   E 50 meters

5. Graham runs in a race that takes him 30 minutes to complete. Which of the following measurements is a likely distance for the race he ran?

   A 5 feet
   B 5 kilometers
   C 5 meters
   D 5 inches
   E 5 centimeters

6. Using the ruler shown, what is the best estimate for the horizontal length of the notecard below?

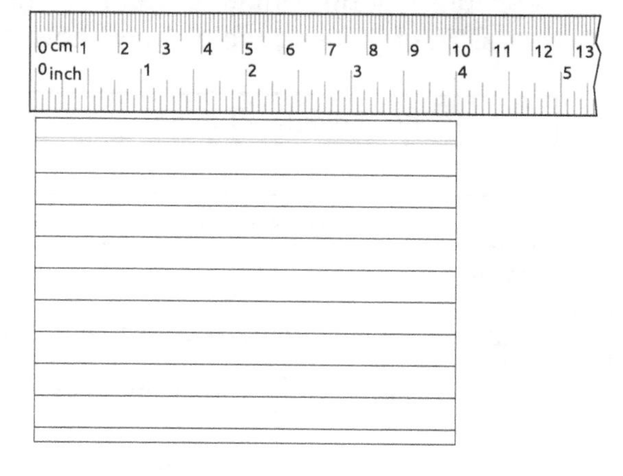

   A 2 inches
   B 2 centimeters
   C 4 inches
   D 4 centimeters
   E 10 inches

7. Which of the following items would be most appropriately measured with a ruler?

   A the length of an Olympic-sized lap pool
   B the width of a plant cell
   C the length of a person's hand
   D the speed of a race car
   E the amount of flour used in a bread recipe

See page 148 for answers and help.

Solving Problems Involving Measurement

Time

It is important to remember that time is a unit of measurement. Some of the basic units of time include seconds, minutes, hours, days, weeks, months, and years. The following table shows how these units relate to each other.

60 seconds = 1 minute	7 days = 1 week
60 minutes = 1 hour	365 days = 1 year*
24 hours = 1 day	12 months = 1 year

*except leap year

Consider the following example using the table to compare units.

Janessa's time card for her work week is shown below.

Monday – 6 hours, 45 minutes
Tuesday – 4 hours, 30 minutes
Wednesday – 7 hours
Thursday – 7 hours, 15 minutes
Friday – 8 hours, 45 minutes

a) How many hours did Janessa work this week?

b) She earns \$14 per hour. How can she calculate her total earnings before taxes for the week?

ANSWER: a) Add the number of minutes: $45 + 30 + 15 + 45 = 135$ minutes. Since there are 60 minutes in 1 hour, 135 minutes is 2 hours and 15 minutes, or 2.25 hours. Add the total hours: $6 + 4 + 7 + 7 + 8 + 2.25 = 34.25$ total hours.

b) Multiply \$14 × 34.25 hours. \$14 × 34.25 = \$479.50

SKILLS TIP

Pay close attention to the units the answer should be in. Remember, each 15 minutes is 0.25 hours.

Liquid Volume

Both the customary and the metric systems of measurement have units that measure liquid volume. The following table shows how many of the common units in both systems relate to each other.

Customary System	Metric System
4 quarts = 1 gallon	1,000 milliliters = 1 liter
32 fluid ounces = 1 quart	1,000 liters = 1 kiloliter
128 fluid ounces = 1 gallon	
2 cups = 1 pint	
8 fluid ounces = 1 cup	

Consider the following example using the table as a reference.

Corey is making a recipe that calls for 2 cups of buttermilk. Another recipe he will also bake calls for 1.5 cups of buttermilk. He started with 1 quart of buttermilk. After finishing the recipes, how many cups of buttermilk does he have left?

ANSWER: Since there are 8 fluid ounces in 1 cup and 32 fluid ounces in 1 quart, there are 4 cups in 1 quart. So, he has used 3.5 cups of the 4 cups he started with. He has 0.5 cups left.

Coach Bell tells his runners that they should drink at least 3.5 liters of water the day before the race. Petra has a water bottle that holds 600 milliliters of fluid. About how many bottles of water does she need to drink the day before the race?

ANSWER: 600 milliliters is 0.6 liter. So, $3.5 \div 0.6 = 5.8$. She should drink about 6 bottles of water.

Mass

Weight and mass can be measured in both the customary and the metric systems of measurement as well. The following table shows how many of the common units in both systems relate to each other.

Customary System	Metric System
2,000 pounds = 1 ton	1,000 grams = 1 kilogram
16 ounces = 1 pound	1,000 milligrams = 1 gram
	1,000 kilograms = 1 metric ton

Use the table as a reference to answer the following question.

The weight limit on a certain elevator is 1.5 tons. If the weight limit is exceeded, an alarm will sound. If 8 people whose average weight is 180 pounds get on the elevator at once, will the alarm sound? Explain.

ANSWER: $180 \times 8 = 1{,}440$ pounds. 1.5 tons is 3,000 pounds. The alarm will not sound.

SKILLS TIP

When changing from smaller units, milliliters, to larger units, liters, divide. Since there are 1,000 milliliters in 1 liter, divide the number of milliliters by 1,000.

SKILLS TIP

When using a formula, it is often helpful to write the formula out as the first step in your work. Then, identify all of the known quantities given to you in the problem and substitute them into the formula.

Solving Problems Involving Measurement

KEY POINT!

6 inches is 0.5 feet. Since each foot is equal to 12 inches, the 6 inches on both ends combine to equal 1 foot.

Rate Problems

A **rate** is a ratio comparing the change in one unit to the change in another unit. A familiar rate used when driving is miles per hour, which compares the distance traveled, in miles, to the time spent driving, in hours. A very important formula is used that relates distance to the rate, or speed, of travel and the time spent traveling. The formula is

$$\text{distance} = \text{rate} \times \text{time}$$

Distance is measured in units of length like miles, feet, kilometers, meters, and others. Time is primarily measured in seconds, minutes, and hours.

Joshua drives at an average speed of 45 miles per hour for 3 hours. How far did he drive?

ANSWER: $d = r \times t$

$d = 45 \times 3$

$d = 135$ miles

Farrah is running in a race. She ran the first 3 miles in 22 minutes and 30 seconds. What is her rate in minutes per mile?

ANSWER: $r = \dfrac{\text{minutes}}{\text{mile}}$

$r = \dfrac{22.5}{3}$

$r = 7.5$

Her rate is 7.5 minutes, or 7 minutes and 30 seconds, per mile.

Complete the activities below to check your understanding of the lesson content.

1. Jimmy, Osman, Kareem, and Jong ran a relay race together. Their times were 52 seconds, 54 seconds, 56 seconds, and 49 seconds, respectively. What was the total of their race times?
   - **A** 331 seconds
   - **B** 2 minutes and 31 seconds
   - **C** 3 minutes and 31 seconds
   - **D** 4 minutes and 31 seconds
   - **E** 211 minutes

2. Neville made 12 liters of punch for a wedding reception. The guests drank 8,500 milliliters of punch. How much punch was left over from the wedding?
   - **A** 350 milliliters
   - **B** 3,500 milliliters
   - **C** 4,500 milliliters
   - **D** 8,488 milliliters
   - **E** 20,500 milliliters

3. A drink requires $\frac{1}{2}$ cup of sugar for each gallon of drink made. Which of the following would give the same concentration of sugar for the drink?
   - **A** 2 cups of sugar for 2 gallons
   - **B** 3 cups of sugar for 3 gallons
   - **C** $\frac{1}{4}$ cup of sugar for $\frac{1}{2}$ gallon
   - **D** $\frac{1}{4}$ cup of sugar for 2 gallons
   - **E** 4 cups of sugar for 2 gallons

Questions 4–6 are based on the following content.

A neighborhood pool is 30 feet long by 15 feet wide and is surrounded by a deck of the same width all the way around the pool.

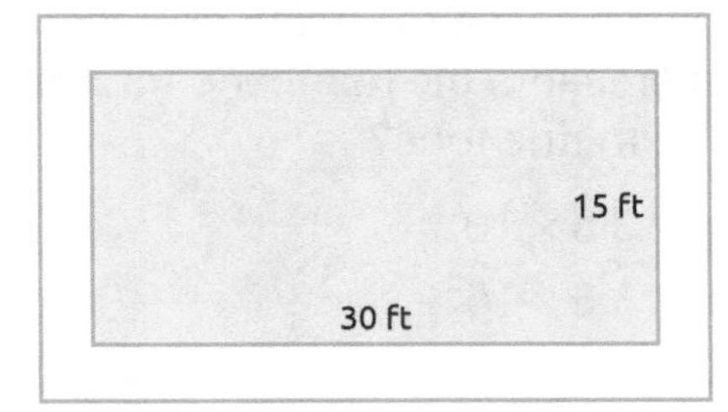

4. The deck goes all the way around the pool and has a width of 3 feet 6 inches. How long is the longest side of the deck?
   - **A** 7 feet
   - **B** 15 feet
   - **C** 30 feet
   - **D** 33 feet 6 inches
   - **E** 37 feet

5. The deck around the pool is 3 feet 6 inches wide. How long is the shortest side of the deck?
   - **A** 7 feet
   - **B** 15 feet
   - **C** 18 feet 6 inches
   - **D** 22 feet
   - **E** 37 feet

6. What is the perimeter of the outside edge of the deck?
   - **A** 90 feet
   - **B** 52 feet
   - **C** 59 feet
   - **D** 118 feet
   - **E** 814 feet

TEST STRATEGY

Pay careful attention to the units listed in the answer choices. Sometimes, the units can help to eliminate answer choices that are unreasonable.

Lesson Practice

7. Use the table to answer the following question.

1,000 milligrams = 1 gram
1,000 grams = 1 kilogram
1,000 kilograms = 1 metric ton

Using the values in the table, which of the following expressions represents the number of grams in 3.5 metric tons?

A 3.5×10^{-6}
B 3.5×10^{-3}
C 3.5×10^{3}
D 3.5×10^{6}
E 3.5×10^{9}

8. A teacher is measuring the height of her students to the nearest $\frac{1}{100}$ of a meter. If a student's height is recorded as 1.80 meters tall, the student's height was actually between which two amounts?

A 1.75 and 1.84 meters
B 1.795 and 1.804 meters
C 1.80 and 1.85 meters
D 1.80 and 1.804 meters
E 1.795 and 1.8 meters

See page 148 for answers and help.

9. A rectangular-shaped walkway through a garden measures 32 feet by 6 feet. The landscape company is going to lay stone pavers along the walkway that each cover 2 square feet. Which of the following expressions can be used to determine the number of paver stones that are needed for the walkway?

A $\frac{32 \times 6}{2}$
B $\frac{2}{32 \times 6}$
C $\frac{32 \times 2}{6}$
D $\frac{2 \times 6}{32}$
E $32 \times 6 \times 2$

Unit Multipliers

One way to convert between different units is to utilize unit multipliers. **Unit multipliers** are fractions worth a value of 1 that can be used to convert quantities to different units. For example, since 3 feet equal 1 yard, you can write a unit multiplier of $\frac{3\text{ feet}}{1\text{ yard}}$ or $\frac{1\text{ yard}}{3\text{ feet}}$. Since the values in both the numerator and denominator are equivalent, both ratios equal 1. You can convert quantities by using unit multipliers and canceling units accordingly. Consider the following example.

Convert 5 yards to feet using unit multipliers.

ANSWER: $\frac{5\text{ }\cancel{\text{yards}}}{1} \times \frac{3\text{ feet}}{1\text{ }\cancel{\text{yards}}} = 15\text{ feet}$

SKILLS TIP

When using unit multipliers correctly, the units will cancel each other out. It the units do not cancel, flip the unit multiplier so that they do.

Length

Some common unit multipliers used for converting various lengths are shown in the following table.

Customary Units	Metric Units
$\frac{3\text{ ft}}{1\text{ yd}}$ or $\frac{1\text{ yd}}{3\text{ ft}}$	$\frac{1{,}000\text{ mm}}{1\text{ m}}$ or $\frac{1\text{ m}}{1{,}000\text{ mm}}$
$\frac{12\text{ in.}}{1\text{ ft}}$ or $\frac{1\text{ ft}}{12\text{ in.}}$	$\frac{100\text{ cm}}{1\text{ m}}$ or $\frac{1\text{ m}}{100\text{ cm}}$
$\frac{5{,}280\text{ ft}}{1\text{ mi}}$ or $\frac{1\text{ mi}}{5{,}280\text{ ft}}$	$\frac{1{,}000\text{ m}}{1\text{ km}}$ or $\frac{1\text{ km}}{1{,}000\text{ m}}$
$\frac{1{,}760\text{ yards}}{1\text{ mile}}$ or $\frac{1\text{ miles}}{1{,}760\text{ yards}}$	$\frac{1\text{ in.}}{2.54\text{ cm}}$ or $\frac{2.54\text{ cm}}{1\text{ in.}}$

Mrs. Hernandez has a model train track in her classroom that has 375 centimeters of track. How many inches of track does she have? Use unit multipliers and round to the nearest inch.

ANSWER: $\frac{375\text{ }\cancel{\text{cm}}}{1} \times \frac{1\text{ in.}}{2.54\text{ }\cancel{\text{cm}}} \approx 148\text{ in.}$

Shanika ran in the 400-meter race. How many kilometers did she run in the race?

ANSWER: $\frac{400\text{ }\cancel{\text{m}}}{1} \times \frac{1\text{ km}}{1{,}000\text{ }\cancel{\text{m}}} = 0.40\text{ km}$

Converting Measurement Units

SKILLS TIP

One advantage to metric units is that they differ from each other by a power of 10. When converting between metric units, simply multiply or divide by the correct power of 10. This can be done by moving the decimal point to the left when dividing and to the right when multiplying.

Liquid Volume and Mass

Liquid volume and mass can be measured using customary and metric units as well. Look at the following table.

Liquid Volume (Customary)	Liquid Volume (Metric)	Mass (Customary)	Mass (Metric)
$\frac{4\text{ cups}}{1\text{ qt}}$ or $\frac{1\text{ qt}}{4\text{ cups}}$	$\frac{1{,}000\text{ mL}}{1\text{ L}}$ or $\frac{1\text{ L}}{1{,}000\text{ mL}}$	$\frac{16\text{ oz}}{1\text{ lb}}$ or $\frac{1\text{ lb}}{16\text{ oz}}$	$\frac{1{,}000\text{ mg}}{1\text{ g}}$ or $\frac{1\text{ g}}{1{,}000\text{ mg}}$
$\frac{2\text{ pints}}{1\text{ qt}}$ or $\frac{1\text{ qt}}{2\text{ pints}}$	$\frac{1{,}000\text{ L}}{1\text{ kL}}$ or $\frac{1\text{ kL}}{1{,}000\text{ L}}$	$\frac{2{,}000\text{ lbs}}{1\text{ ton}}$ or $\frac{1\text{ ton}}{2{,}000\text{ lbs}}$	$\frac{1{,}000\text{ g}}{1\text{ kg}}$ or $\frac{1\text{ kg}}{1{,}000\text{ g}}$
$\frac{4\text{ qt}}{1\text{ gal}}$ or $\frac{1\text{ gal}}{4\text{ qt}}$			$\frac{1{,}000\text{ kg}}{1\text{ metric ton}}$ or $\frac{1\text{ metric ton}}{1{,}000\text{ kg}}$

In a recipe, each gallon of punch calls for 1 quart of orange juice. Jasmine has 2 gallons of orange juice. How many gallons of punch can she make with her orange juice?

ANSWER: Convert the 2 gallons of juice to quarts: $\frac{2\cancel{\text{ gal}}}{1} \times \frac{4\text{ qt}}{1\cancel{\text{ gal}}} = 8\text{ qt}$ Since each gallon calls for 1 quart, she has enough orange juice to make 8 gallons of punch.

Complete the activities below to check your understanding of the lesson content.

1. Ethan is constructing a shed. He measures and finds the length of the shed to be 222 inches. Which of the following represents the length in feet?
   - **A** 22 feet
   - **B** 19 feet
   - **C** 18.5 feet
   - **D** 18 feet
   - **E** 17.5 feet

2. Diego placed a bucket below a leaky water line. The bucket can hold 5 liters of water. The drip is filling the bucket at a rate of 125 milliliters per hour. At this rate, how long will it take for the bucket to be filled?
   - **A** 40 minutes
   - **B** 40 hours
   - **C** 25 hours
   - **D** 625 minutes
   - **E** 625 hours

3. Salmina and her sister, Abigail, are both training for a race. Salmina runs a course that is 1.5 miles long. Abigail runs a course that is 9,240 feet long. Who runs farther? Explain how you know.

   ____

   ____

   ____

4. The football team needs 40 more yards to score a touchdown. On the next play, the running back ran for 36 yards. The following play, he ran for 5 feet. How many more inches does he need to run to score a touchdown?
   - **A** 1 inch
   - **B** 7 inches
   - **C** 12 inches
   - **D** 84 inches
   - **E** 91 inches

5. Keila is running a race that is 10 kilometers long. If the track is 400 meters long, how many laps does Keila need to run to complete the race?
   - **A** 10
   - **B** 15
   - **C** 25
   - **D** 30
   - **E** 400

6. The food charity sets a goal of collecting 6 gallons of soup in October. So far, they have collected 50 pints of soup. Have they met their goal? Explain.

   ____

   ____

   ____

KEY POINT!

When converting from smaller units, like inches, to larger units, like feet, divide. Or, use a unit multiplier: $\frac{222\text{ in.}}{1} \times \frac{1\text{ ft}}{12\text{ in.}}$

TEST STRATEGY

Read the problem carefully, paying close attention to the given units.

Lesson Practice

7. The class has planted a sunflower and is charting its growth each week. On average, the plant has grown 40 centimeters each week. At this rate, how long will it take for the sunflower to grow to be 3.2 meters tall?
   - **A** 3.2 weeks
   - **B** 4 weeks
   - **C** 6 weeks
   - **D** 8 weeks
   - **E** 12 weeks

8. James lives 5.5 miles from school. Mrs. Harris drives him to her home, which gets James 4.8 miles closer to home. He then walks the remaining distance. How far does James walk?
   - **A** 5,280 feet
   - **B** 4,224 feet
   - **C** 3,696 feet
   - **D** 1,408 feet
   - **E** 0.8 feet

See page 148 for answers and help.

Answer the questions based on the content covered in this unit.

1. Greta is going to run sprints on the straightaways of her high school track for her workout. Which of the following is a reasonable length for the straightaway?
   - **A** 40 inches
   - **B** 4 feet
   - **C** 4 miles
   - **D** 4 kilometers
   - **E** 40 meters

2. Johnson is running a race that is 5 kilometers long. If the track is 400 meters long, how many laps does Johnson need to run to complete the race?
   - **A** 1.25
   - **B** 12.5
   - **C** 25
   - **D** 30
   - **E** 400

Questions 3–5 refer to the following information.

DePaul is designing a two-colored patio that has a red, rectangular-shaped area inside that is 18 meters long by 8 meters wide. The red is surrounded by a gray border that is the same width all the way around.

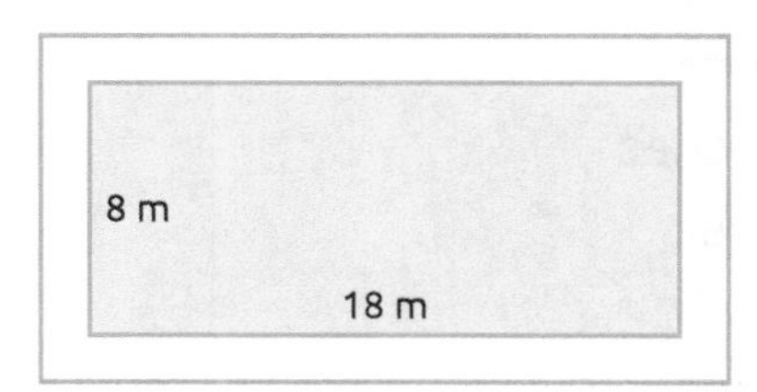

3. The gray border extends beyond the red rectangle by 2 meters 25 centimeters on each side. If the length of the red rectangle is 18 meters, what is the length of the gray border?
   - **A** 18 meters
   - **B** 20 meters 25 centimeters
   - **C** 22 meters 25 centimeters
   - **D** 22 meters 50 centimeters
   - **E** 22 meters 75 centimeters

4. The gray border extends beyond the red region by 2 meters 25 centimeters on each side. If the width of the red rectangle is 8 meters, what is the width of the gray border?
   - **A** 8 meters
   - **B** 10 meters 25 centimeters
   - **C** 12 meters 50 centimeters
   - **D** 12 meters 75 centimeters
   - **E** 14 meters 50 centimeters

5. What is the perimeter of the outside edge of the gray border?
   - **A** 70 meters
   - **B** 61 meters
   - **C** 52 meters
   - **D** 30 meters 50 centimeters
   - **E** 26 meters

6. Zusi drives to visit her friend at college, and it takes her 4 hours to get there. Which of the following distances is most likely the distance she drove?
   - **A** 10 miles
   - **B** 20 miles
   - **C** 50 miles
   - **D** 200 miles
   - **E** 1,000 miles

7. Jarvin has a roll of yarn 15 feet long. He measures and cuts off 8 pieces that are each 16 inches long. How much yarn does he have left in his roll?
   - **A** $3\frac{1}{2}$ feet
   - **B** $4\frac{1}{2}$ feet
   - **C** $4\frac{1}{3}$ feet
   - **D** $10\frac{1}{2}$ feet
   - **E** $10\frac{2}{3}$ feet

8. Using the ruler shown, what is the best estimate for the length of the caterpillar in the picture?

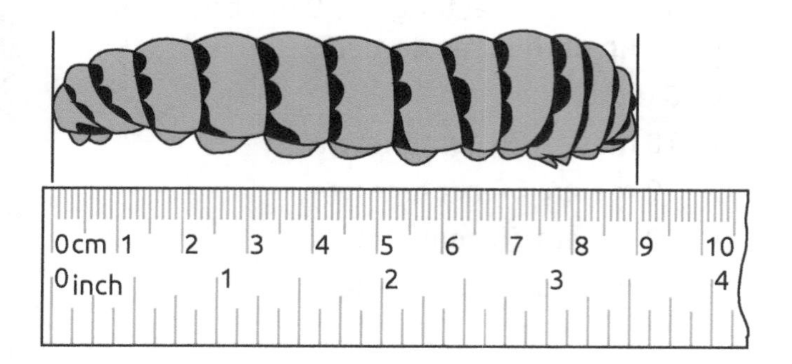

A 2 inches
B 3.5 centimeters
C 9 inches
D 9 centimeters
E 12 inches

9. Which of the following items would be appropriate to measure with a ruler?
A a race track for motorcycles
B a sheet of notebook paper
C the length of a classroom
D the amount of water a jug can hold
E the amount of time muffins need to bake

10. Sam has four boards of varying lengths. The lengths are 6 feet 8 inches, 5 feet 11 inches, 7 feet 2 inches, and 6 feet 3 inches. What is the total length of all four boards?
A 24 feet
B 25 feet
C 26 feet
D 25 feet 10 inches
E 26 feet 2 inches

Questions 11–13 refer to the drawing of a playground below.

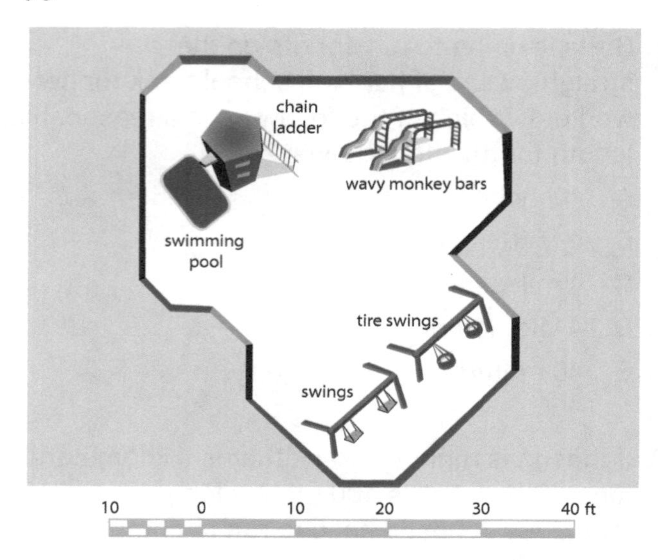

11. In the drawing, the swing set is said to be 25 units long. What units are most likely being referred to?
A centimeters
B kilometers
C yards
D inches
E feet

12. In the drawing, what is the most likely width of the monkey bars?
A 5 feet
B 6 inches
C 5 yards
D 18 inches
E 20 feet

13. ABC Landscape is going to lay mulch throughout the play area. Which of the following instruments would be the best choice to determine the measurements of the playground?
A foot-long ruler
B 50-foot tape measure
C yardstick
D meterstick
E mile marker

14. A hole on a golf course is 102 yards long. Felipe hit the golf ball 88 yards off the tee toward the hole. He then chipped the ball 38 feet closer to the hole. How many inches is Felipe away from the hole?
    - **A** 4 inches
    - **B** 8 inches
    - **C** 12 inches
    - **D** 24 inches
    - **E** 48 inches

15. A drink requires $2\frac{1}{2}$ grams of powder mix for each liter of drink made. Which of the following would give the same concentration of powder mix for the drink?
    - **A** 5 grams for 2 liters
    - **B** 6 grams for 3 liters
    - **C** $1\frac{3}{4}$ grams for $\frac{1}{2}$ liter
    - **D** $3\frac{1}{2}$ grams for 3 liters
    - **E** 10 grams for 10 liters

Use this table to answer the following question.

1,000 millimeters = 1 meter
1,000 meters = 1 kilometer
1,000 kilometers = 1 megameter

16. Using the values in the table, which of the following expressions represents the number of meters in 5.6 megameters?
    - **A** 5.6×10^{9}
    - **B** 5.6×10^{6}
    - **C** 5.6×10^{3}
    - **D** 5.6×10^{-3}
    - **E** 5.6×10^{-6}

17. A tub can hold 30 liters of water. It is filling at a rate of 600 milliliters per hour. At this rate, how long will it take for the tub to be filled?
    - **A** 5 minutes
    - **B** 5 hours
    - **C** 50 minutes
    - **D** 50 hours
    - **E** 500 hours

18. At practice, a coach is recording her runners' times to the nearest $\frac{1}{100}$ of a second. If a runner's time is recorded as 5.70 seconds, the runner's time was actually between which two amounts?
    - **A** 5.65 and 5.74 seconds
    - **B** 5.695 and 5.7 seconds
    - **C** 5.60 and 5.75 seconds
    - **D** 5.60 and 5.704 seconds
    - **E** 5.695 and 5.704 seconds

19. A rectangular-shaped roof measures 168 feet by 72 feet. A roofing company is going to use slate shingles that each cover 3 square feet. Which of the following expressions can be used to determine the number of slate shingles that are needed for the roof?
    - **A** $\frac{168 \times 3}{72}$
    - **B** $\frac{3}{168 \times 72}$
    - **C** $\frac{168 \times 72}{3}$
    - **D** $\frac{3 \times 168}{72}$
    - **E** $168 \times 72 \times 3$

See page 148 for answers.

Lesson 1

1. **E.** Feet are the most likely units for the closet. All other options are either much too short or long.
2. B.
3. **C.** The tape measure allows for measurement of each wall continuously. The other instruments are not long enough to measure the walls without stacking them end on end many times.
4. **D.** 50 feet is the most reasonable answer choice for an indoor facility. The other choices are much too short or much too long.
5. **B.** Kilometers is the only one long enough. All of the other options are too short in terms of distance.
6. **C.** According to the ruler, the horizontal length of the card is about 4 inches.
7. C.

Lesson 2

1. **C.** The sum of the four times is 211 seconds. Convert to minutes and seconds.
2. **B.** 12 liters is 12,000 milliliters. Subtract to find the amount left over.
3. C.
4. **E.** The deck extends beyond the length of the pool by 3.5 feet in both directions. The length is $30 + 3.5 + 3.5$ feet.
5. D.
6. **D.** Add 7 feet to the length and the width of the pool. Add twice the length and twice the width to find the perimeter of the deck.
7. **D.** There are 1,000 kilograms in a metric ton and 1,000 grams in 1 kilogram. Multiply $3.5 \times 1{,}000 \times 1{,}000$.
8. B.
9. A.

Lesson 3

1. **C.** There are 12 inches in a foot. Divide 222 by 12.
2. **B.** Convert 5 liters to milliliters. Then, divide 5,000 mL by 125 mL.
3. Use unit multipliers to convert one of the measurements so they are the same. $\frac{1.5 \text{ mi}}{1} \times \frac{5{,}280 \text{ ft}}{1 \text{ mi}} = 7{,}920 \text{ ft}$ or $\frac{9{,}240 \text{ ft}}{1} \times \frac{1 \text{ mi}}{5{,}280 \text{ ft}} = 1.75 \text{ mi}$. Then compare. Abigail runs farther.
4. **D.** After the first run, they still need to go 4 yards, or 12 feet. After running 5 more feet, they have 7 more feet to go, which is 84 inches.
5. C.
6. Convert 6 gallons to pints: $\frac{6 \text{ gal}}{1} \times \frac{8 \text{ pt}}{1 \text{gal}} = 48 \text{ pt}$ Yes, they have met their goal. They needed 48 pints to meet their goal, and they received 50.
7. D.
8. **C.** Subtract 5.5 – 4.8 to find that he must walk 0.7 miles. Then, convert miles to feet.

Unit Test

1. E.
2. B.
3. D.
4. C.
5. A.
6. D.
7. C.
8. D.
9. B.
10. C.
11. E.
12. D.
13. B.
14. E.
15. A.
16. B.
17. D.
18. E.
19. C.

Unit Glossary

UNIT 6

- **English units of measurement** – units utilizing inches, feet, yards, and miles for length; ounces and pounds for mass; and fluid ounces, quarts, and gallons for liquid measurement
- **Metric units of measurement** – units utilizing meters for length; grams for mass; and liters for liquid measurement
- **rate** – a ratio of change comparing the change in one unit to the change in another unit
- **unit multiplier** – fractions worth a value of 1 that can be used to convert quantities to different units

Study More!

UNIT 8

Measure and Estimate Lengths Involving Standard Units

- Know how to measure and estimate other types of measurement, like mass, time, and liquid measurements.
- Combine length measurements and estimations with formulas for perimeter and area of rectangles and triangles.

Solve Problems Involving Rate

- Know how to work with various rates and convert from one rate to another.
- Know how to solve more advanced rate problems.
- Know how to solve word problems involving rate.

Convert Measurement Units

- Know how to convert between customary and metric units.
- Know how to convert between square units and cubic units.

UNIT 7

Statistics and Probability

Student	Score (%)
Faisal	92
Jira	78
Kelley	89
Cory	81
Fernando	70
Katriella	94
Lucy	88

Statistics and probability have many applications in life as we work to study and understand data. Data can be used to summarize things that have happened and make predictions based on what is thought to be normal occurrences. Consider the following scenario:

> Mrs. Diaz is giving a math test and hopes that the average score will be at least 83%. The results are shown in the following table. Did the class achieve Mrs. Diaz's goal?
>
> **Answer:** Yes, the class achieved the goal.

KEY WORDS

- bias
- box-and-whisker plot
- circle graph
- event
- frequency
- Fundamental Counting Principle
- histogram
- inference
- interquartile range
- mean
- measures of central tendency
- median
- mode
- outcome
- outlier
- population
- probability
- quartile
- random sample
- range
- sample
- scatter plot
- tree diagram

Measures of Center

The **measures of central tendency** are numbers that describe the center of a set of data. These measures are summarized in the table below.

Mean	The sum of the numbers in a data set divided by the number of items in the set
Median	The middle number of a set arranged from least to greatest, or the average of the two middle numbers if the data set contains an even number of numbers
Mode	The number or numbers that appear most often in the data set
Range	The difference between the greatest number and the least number in the data set

Typically, mean, median, and mode are all different ways of finding the measure of center. The range measures the spread of the data. Range is simply how spread out the data is from the least to the greatest value in the data set.

During the first half of the 2014 regular season, the Dallas Cowboys scored the following numbers of points in each game: 17, 26, 34, 38, 20, 30, 31, and 17. What are the mean, median, mode, and range of their scores through the first half of the season?

ANSWER: Mean: $\frac{17+26+34+38+20+30+31+17}{8}=\frac{213}{8}\approx 26.6$ points per game

Median: 17, 17, 20, (26, 30) 31, 34, 38 $\frac{26+30}{2}=\frac{56}{2}=28$

Mode: 17

Range: $38-17=21$

Measures of center can also be analyzed to give valuable information about a data set. The methods for determining measures of center can be useful when describing the entire data set.

At a used car lot, the mean price for a car is $2,400. What things can be determined about the prices for all the cars on the lot?

ANSWER: The mean value of $2,400 is likely somewhere in the middle of a list of the prices for all the cars. It is most likely that some cars in the sample will cost more than $2,400 and some will cost less than $2,400. The mean value is always between the least and greatest values, so the least expensive car on the lot is $2,400 or less, and the most expensive car on the lot is $2,400 or more.

Box-and-Whisker Plots

A **box-and-whisker plot** is a graph that shows how spread out a set of data is. It uses a number line and a box, which shows the median of the data, and the first and third quartiles of the data.

Quartiles are three numbers that divide the data into four equal parts. The median is the 2nd quartile. To find the 1st quartile, first identify the lower half of the data set, or all the numbers less than the median. The 1st quartile is the median of the lower half of the data set. To find the 3rd quartile, identify the upper half of the data set, or all the numbers greater than the median. The 3rd quartile is the median of the upper half of the data set.

SKILLS TIP

The measure of central tendency that best represents the middle of the data will vary for different data sets. It is important to consider the data as a whole, and analyze the measures to see what each indicates about the data.

Understanding Statistical Variability

The **interquartile range** is the range between the 1st and the 3rd quartiles, or the length of the box. The whiskers extend beyond the box to show the minimum and the maximum values of the data set. **Outliers** are data points that are much larger or much smaller than most of the values in the set. Outliers can skew the data and sometimes are excluded from the data when determining measures of center and spread.

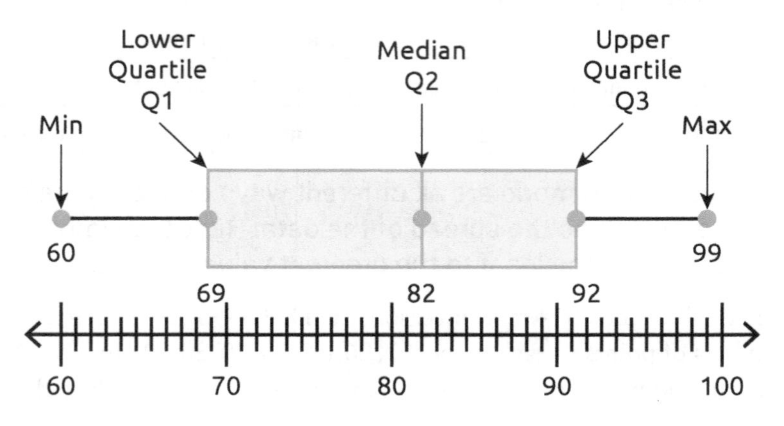

Draw a box-and-whisker plot that represents the number of wins the Dallas Cowboys had in the first half of the season. Follow these steps to create a box-and-whisker plot.

1) Draw a number line that includes the minimum, median, and maximum numbers of the data set.

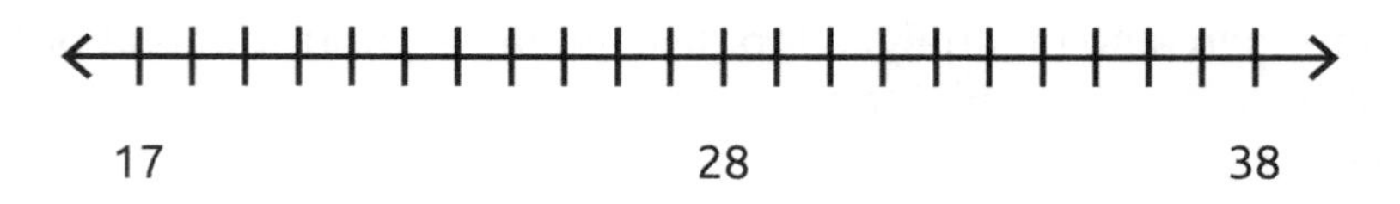

2) Mark the minimum, median, maximum, and the 1st and 3rd quartiles on the number line.

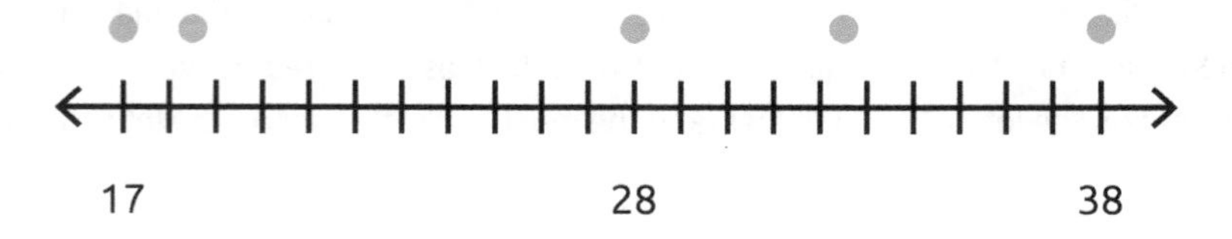

3) Draw the box and the whiskers.

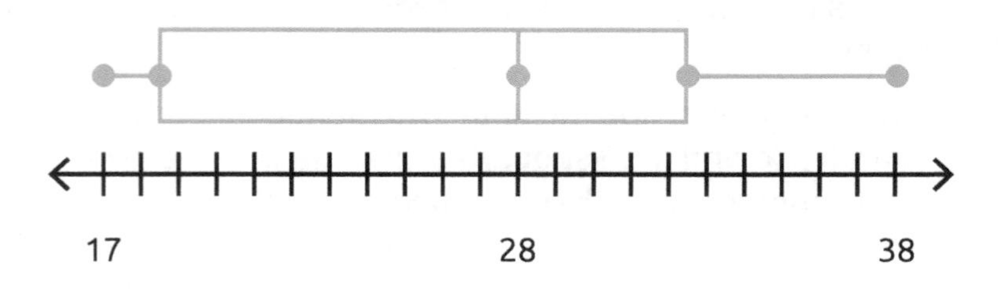

SKILLS TIP

To find the 1st and 3rd quartiles, find the median first. The 1st quartile is the median of the lower half of the values. The 3rd quartile is the median of the upper half of the values.

Complete the activities below to check your understanding of the lesson content.

The seating capacity at several U.S. soccer-specific stadiums is shown in the table below. Questions 1–4 refer to the following table.

Stadium	Seating Capacity
BBVA Compass Stadium (Houston, TX)	22,039
Avaya Stadium (San Jose, CA)	18,000
Dick's Sporting Goods Park (Commerce City, CO)	18,086
Mapfre Stadium (Columbus, OH)	20,145
Stubhub Center (Carson, CA)	27,000

1. What is the mean seating capacity for these five stadiums?
   A 9,000
   B 20,145
   C 21,054
   D 21,155
   E 27,000

2. What is the median seating capacity for these five stadiums?
   A 9,000
   B 20,145
   C 21,054
   D 21,155
   E 27,000

3. What is the seating capacity range for these five stadiums?
   A 9,000
   B 20,145
   C 21,054
   D 21,155
   E 27,000

4. The Stubhub Center is considering changing their seating capacity. The change would decrease the mean seating capacity for all 5 stadiums. Describe the change Stubhub Center is considering. Explain your reasoning.

   ____

   ____

   ____

TEST STRATEGY

Answer choices often contain unreasonable answers that can be eliminated simply by reading the question and the answer choices carefully.

Lesson Practice

KEY POINT!

An outlier lies significantly outside the range of the rest of the data. Outliers usually affect mean and range the most.

SKILLS TIP

To find the 1st and 3rd quartiles, find the median first. The 1st quartile is the median of the lower half of the values. The 3rd quartile is the median of the upper half of the values.

Questions 5 and 6 refer to the following diagram.

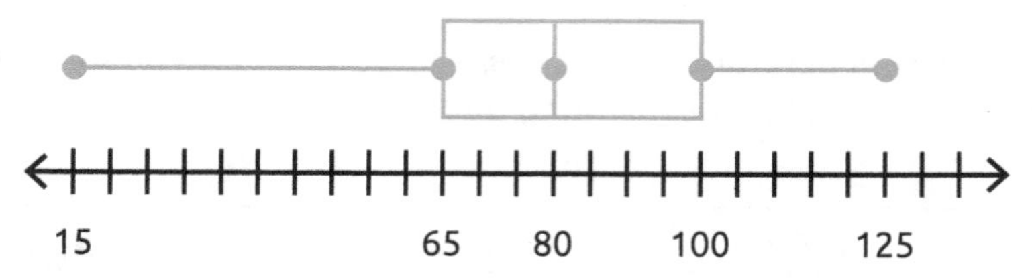

5. What does 15 represent on the box-and-whisker plot?
   A 1st quartile
   B median
   C 3rd quartile
   D maximum
   E outlier

6. Which of the following is NOT true?
   A About half of the data is above 80.
   B All of the data is greater than or equal to 15.
   C The mean is the best measure of center.
   D About one-fourth of the data is greater than 100.
   E About one-fourth of the data is less than 65.

See page 171 for answers and help.

Random Sampling

When someone wants to understand the opinions of a large group of people, a method called sampling can be useful. **Sampling** is a way to select a portion of the larger group, called the **population**. A **random sample** is a portion of the population that is selected in such a way that it represents the entire population. Random sampling can help you to make inferences about a population. An **inference** is a conclusion that you draw about a population based on evidence gathered from a random sample.

For example, in a school with 1,000 students, the administration would like to know what type of hot lunch the student body likes best. They survey a random sample of students by randomly choosing 50 students who represent the whole school. The results of the survey are shown in the following table.

Favorite Lunch	Number of Students
Pizza	28
Spaghetti	8
Sub Sandwich	10
Turkey and Noodles	4

The administration can use the results of this survey to make inferences about the entire student body.

Based on the results of the survey, how many students in the entire school would likely choose pizza as their favorite hot lunch?

ANSWER: First, use the numbers in the table to find the percentage of students out of 50 who chose pizza as their favorite.

$$\frac{28}{50} = 56\%$$

About 56% of the student body would likely choose pizza. 56% of 1,000 students is 560. So, 560 students would likely choose pizza as their favorite hot lunch.

Bias

When taking a random sample, be careful of bias. **Bias** occurs when one or more parts of a survey or sample are favored over others. For example, if you are trying to determine the favorite country star of a certain population, polling people as they leave a country music concert would most likely cause the results to be biased toward the singer who performed at the concert. It might be better to conduct the survey at a music store that sells all varieties of country music. In order for a sample to be unbiased, the sample must be random and representative of the population at large.

SKILLS TIP

Proportions can also be used to make inferences about the data in the sample.

$$\frac{\text{Pizza}}{\text{Total}} = \frac{\text{Pizza}}{\text{Total}}$$
$$\frac{28}{50} = \frac{x}{1{,}000}$$
$$50x = 28{,}000$$
$$x = 560 \text{ students like pizza best}$$

SKILLS TIP

By taking a random sample of a population, you can draw conclusions about the characteristics of the entire population. You can also make predictions based on those conclusions. If a sample is biased toward a particular segment of the population, that will affect the validity of your conclusions and predictions.

Using Random Sampling to Draw Inferences

Examine both of the following scenarios and determine if the sample is biased or unbiased. Explain. If the sample is biased, describe a change that can be made to make it unbiased.

a) A manufacturing facility produces fence posts. Every 500th post is pulled from the line to check quality and length.

ANSWER: This sample is unbiased because the sample is selected systematically and randomly at specified intervals.

b) A newspaper conducted a poll at a senior citizens' recreation center to see how people in the community feel about the mayoral candidate.

ANSWER: The survey is biased because it takes place at a location where everyone is from one segment of the population—senior citizens. The newspaper needs to find a location where they can randomly choose people from the community who represent all segments of the population, like in a shopping mall.

UNIT 7 / LESSON 2

Lesson Practice

TEST STRATEGY

On each bar, write the number of people represented for each category. This may take a minute, but it will save time and help you to answer the questions correctly.

Complete the activities below to check your understanding of the lesson content.

Questions 1–4 refer to the following information.

A pet store took a random sample to determine the types of animals pet owners have as pets. The results are shown in the following graph.

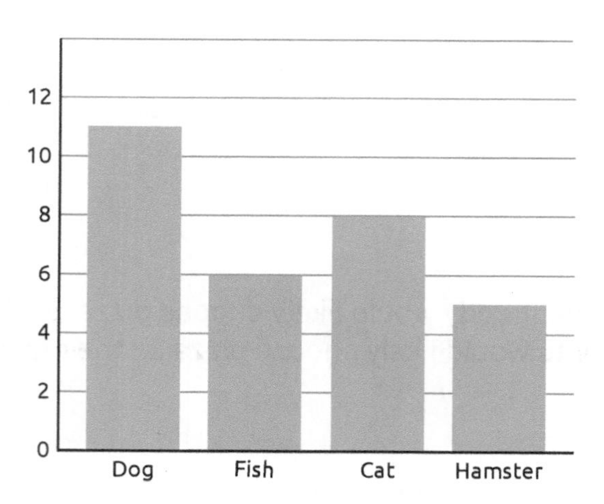

1. How many people are included in the sample?

   A 8
   B 11
   C 12
   D 30
   E 42

2. According to the sample, about what percentage of pet owners own a dog?

   A 50%
   B 37%
   C 30%
   D 20%
   E 11%

3. According to the sample, if the store has 500 pet owners who will visit the store next week to buy food for their pet, about how many will buy cat food?
   - **A** 8
   - **B** 27
   - **C** 135
   - **D** 212
   - **E** 365

4. If the store creates the sample by randomly surveying every 20th customer who comes into their store, is the survey biased or unbiased? Explain your reasoning.

   ____

   ____

   ____

Questions 5–7 refer to the following information.

The community youth recreation center is hosting a celebration in which they are inviting all 700 members. They are going to have ice cream. They survey a sample of members to find out their favorite flavors of ice cream. The results are shown in the following table.

Favorite Flavor	Number of Members
Vanilla	18
Chocolate	13
Strawberry	7
Mint Chocolate Chip	12

5. How many people are included in the sample?
   - **A** 18
   - **B** 25
   - **C** 50
   - **D** 70
   - **E** 700

6. According to the sample, about what percentage of people prefer vanilla?
   - **A** 18%
   - **B** 25%
   - **C** 30%
   - **D** 36%
   - **E** 50%

7. According to the sample, how many of the members at the celebration will want strawberry ice cream?
   - **A** 7
   - **B** 14
   - **C** 70
   - **D** 88
   - **E** 98

See page 171 for answers and help.

KEY POINT!

When you look at a sample, first determine how many people are included.

Describing Distributions

Tables

Tables are useful for summarizing and examining relationships between two categorical variables.

For example, for a data set of 200, comparing people who have blue eyes and blonde hair, the table might be set up as shown below. The **frequency** counts, or the number of times each data type occurs, are listed in the interior cells of the table.

	Blue Eyes	Not Blue Eyes	Total
Blonde Hair	36	64	100
Not Blonde Hair	48	52	100
Total	84	116	200

This frequency table can be used to quickly summarize data about a group of people and whether they have blue eyes or blonde hair. How many people have blue eyes? How many people have neither blue eyes nor blonde hair?

ANSWER: 84 people have blue eyes. 52 people have neither blue eyes nor blonde hair.

SKILLS TIP

Tables can be used to organize a sample involving two variables. The table can then be used to make inferences about the population.

Scatter Plots

A **scatter plot** is a type of graph that can be used to demonstrate the relationship between two sets of data that are graphed as ordered pairs on a coordinate plane. Scatter plots can show a positive relationship, a negative relationship, or no relationship at all. Examples are shown in the table below.

Positive Relationship	Negative Relationship	No Relationship
As x increases, y increases.	As x increases, y decreases.	There is no obvious pattern.

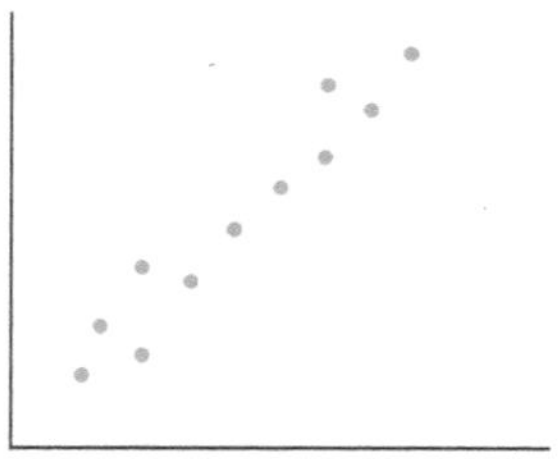

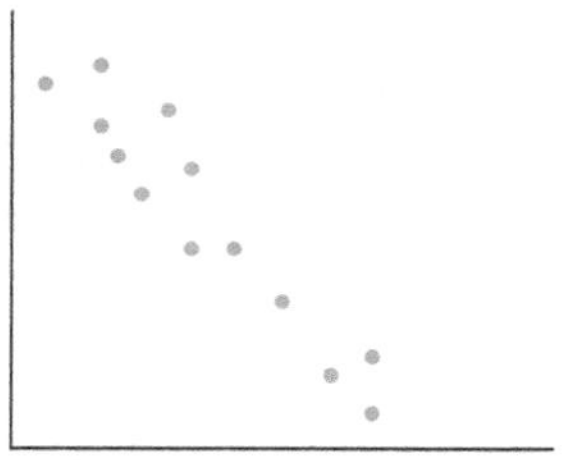

Scatter plots that show a positive or negative relationship can be represented by a line of best fit, which is a line that passes through the middle of a set of data points on a scatter plot. A line of best fit should be close to many of the points, but it doesn't have to pass through any of them. The scatter plot below shows a line of best fit.

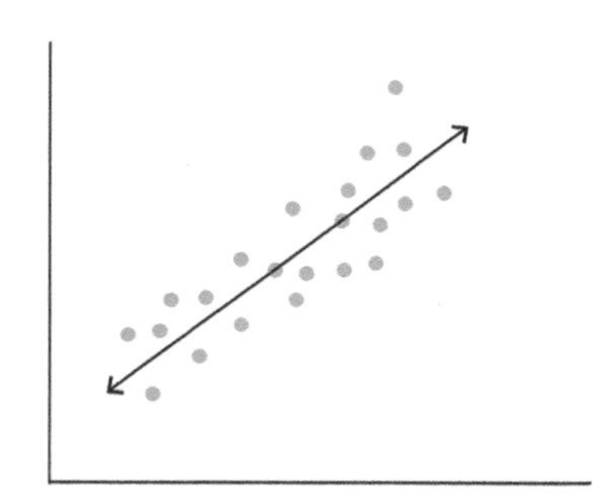

SKILLS TIP

The relationships, both positive and negative, can be described as strong, moderate, or weak, depending on how tightly grouped the data points are. A strong relationship is indicated by a tight grouping and a weak relationship by a looser and more spread out grouping.

Look at the following scatter plot. What is the relationship between time and temperature? As time increases, what happens to the temperature?

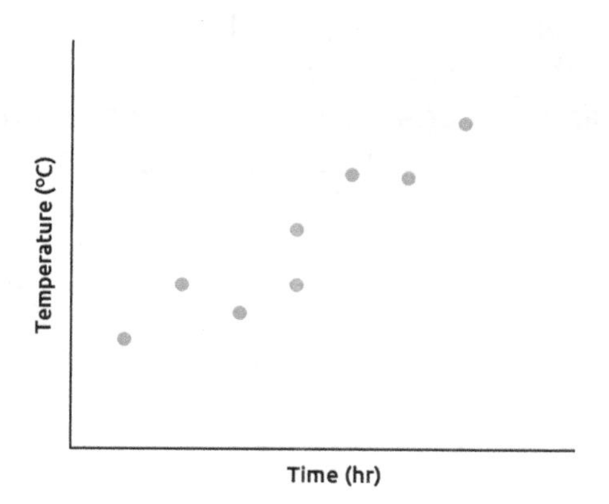

ANSWER: The scatter plot shows a positive relationship. As time increases, temperature increases also.

Histogram

A **histogram** is a graph that displays numerical data that can be organized into equal intervals. Histograms should include a title and labels for the vertical and horizontal axes. The intervals are labeled on the horizontal axis and the frequency along the vertical axis. Consider the following histogram.

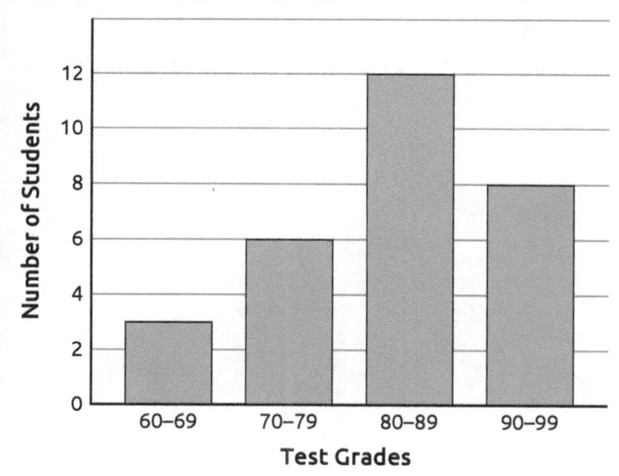

The histogram shows the results of a recent math test that Mr. Jones gave to his class. How many students scored at least 80%?

ANSWER: The histogram shows 12 who scored between 80% and 89% and 8 who scored between 90% and 99%. So, there were 20 students who scored at least 80%.

Circle Graph

A **circle graph** is a graph that displays numerical data as sections of a circle which each section represents a percentage of a total. Large sections of a circle graph represent a large percentage of the total, while small sections represent a small percentage of the total. Circle graphs are useful for comparing different parts of a data set to the entire data set as a whole, like in the circle graph below.

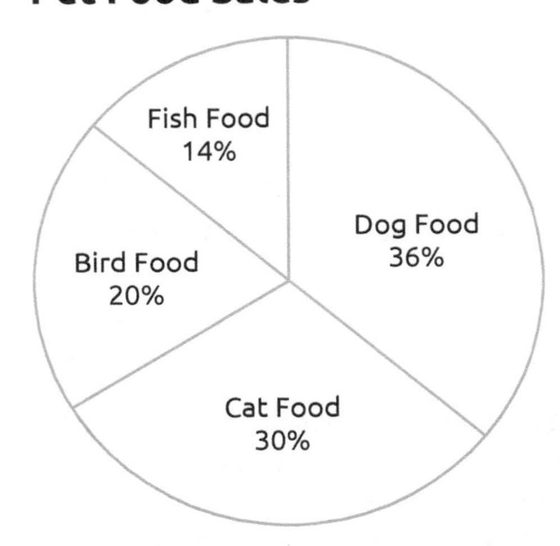

The circle graph shows the monthly food sales for a pet supply store. If $2,500 of pet food was sold, how much fish food was sold?

ANSWER: To find the amount fish food sold, determine 14% of the total, $2,500. (2,500)(0.14) = 350, so $350 of fish food was sold.

Complete the activities below to check your understanding of the lesson content.

Questions 1–3 is based on the following information.

A marketing company surveyed a sample of men and women ages 25–40 to find out how many hours they spend watching sports on TV.

	3 or More Hours per Week	Less than 3 Hours per Week	Total
Men	61	14	75
Women	26	49	75
Total	87	63	150

1. How many people are included in the sample?
   A 61
   B 63
   C 75
   D 87
   E 150

2. According to the table, how many women watch sports for 3 or more hours per week?
   A 26
   B 49
   C 75
   D 87
   E 150

3. According to the table, how many men were surveyed?
   A 26
   B 49
   C 75
   D 87
   E 150

4. The scatter plot shows how practice time affects golf scores. What type of relationship is shown? Explain what that means.

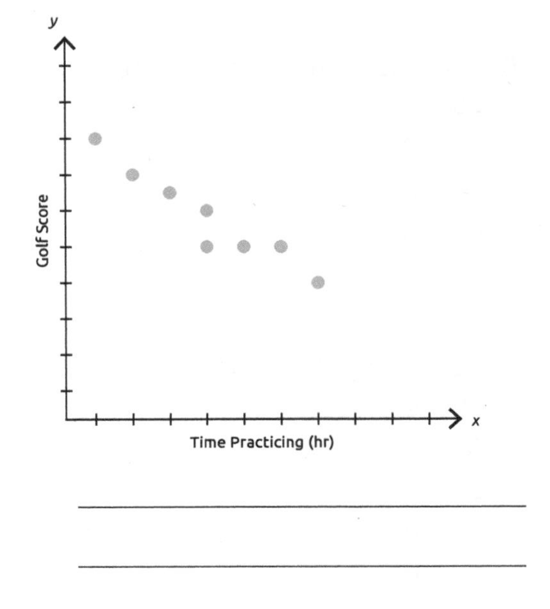

____

____

____

TEST STRATEGY

Read the question carefully, all the way through, once. Then, read the question again, paying close attention to any key words.

Lesson Practice

KEY POINT!

A histogram displays data that occurs in various intervals, which are labeled along the horizontal axis.

Questions 5–7 refer to the following information.

The following histogram shows the total number of wins for all of the teams in the NBA during the 2014–2015 regular season.

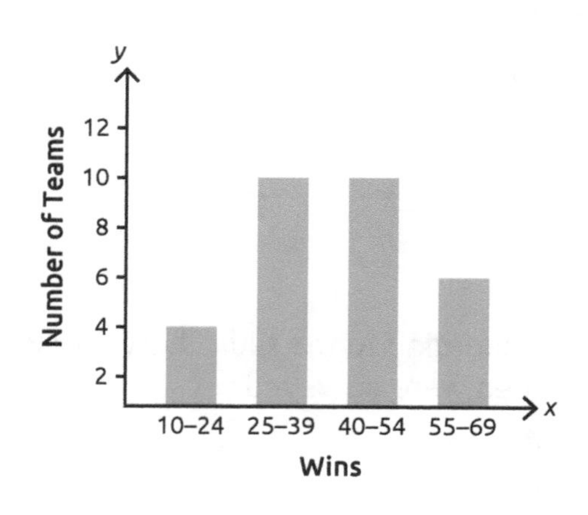

5. How many teams played in the NBA during the 2014–15 regular season?
   - **A** 4
   - **B** 6
   - **C** 10
   - **D** 20
   - **E** 30

6. How many teams won more than 39 games?
   - **A** 20
   - **B** 16
   - **C** 10
   - **D** 6
   - **E** 4%

7. How many teams had 24 wins or fewer?
   - **A** 4
   - **B** 6
   - **C** 10
   - **D** 20
   - **E** 26

A new museum records data on the people who visit. The circle graph shows the percentage of adults, students, and seniors who visited on opening day. Questions 8 and 9 refer to the information below.

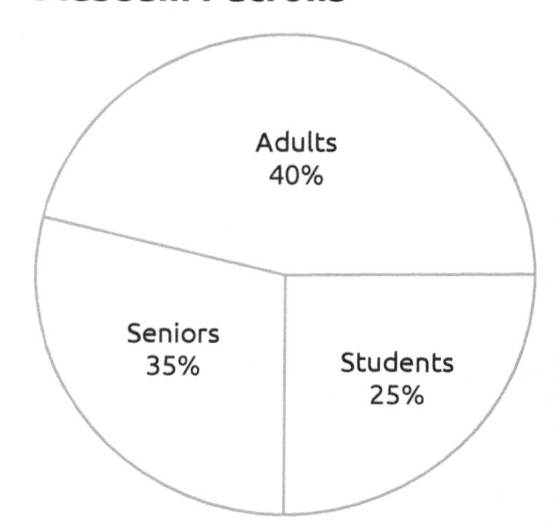

8. If 400 people visited the museum on opening day, how many of the visitors were adults?
   - **A** 10
   - **B** 40
   - **C** 100
   - **D** 160
   - **E** 240

9. The museum expects twice as many visitors the following day. If the percentages remain the same as on opening day, how many seniors should the museum expect?
   - **A** 140
   - **B** 160
   - **C** 280
   - **D** 320
   - **E** 360

See page 171 answers and help.

Tree Diagram

An **outcome** is any of the possible results from a given action. When flipping a coin, two outcomes are possible: heads or tails. An **event** is a specific outcome or outcomes. When multiple outcomes are possible, a **tree diagram** is a way to organize all of the possible outcomes.

A sandwich shop allows customers to choose either white or wheat bread. Customers can also choose turkey, ham, or roast beef. Then they can choose cheddar, provolone, or Swiss cheeses. Use a tree diagram to determine how many possible different sandwich combinations the customers may order.

ANSWER:

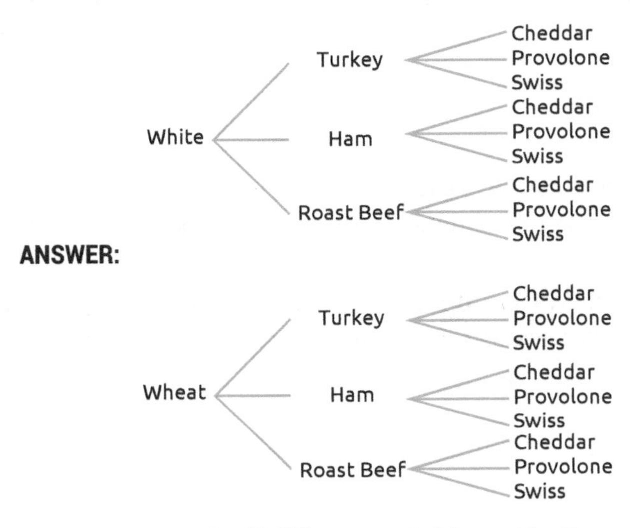

Customers can order 18 different sandwich combinations.

SKILLS TIP

When drawing a tree diagram, start with the initial choices, and then determine how many choices are at the next level of the tree. Repeat until all of the choices are figured into the diagram.

Counting Outcomes

The **Fundamental Counting Principle** gives a way to determine the total number of outcomes possible by multiplying. It says that if event *A* has a possible outcomes and event *B* has b possible outcomes, then event *A* followed by event *B* has $a \cdot b$ total possible outcomes.

Use the Fundamental Counting Principle to determine the number of sandwich combinations customers can make in the previous example.

ANSWER: Since the sandwich shop offers 2 types of bread, 3 types of meat, and 3 types of cheese, the number of outcomes can be determined by multiplying $2 \cdot 3 \cdot 3 = 18$. The shop offers 18 different sandwich combinations.

A combination lock has 3 digits, each of which can be any number between 0 and 9. How many possible lock combinations are there?

ANSWER: 10 numbers are possible for each digit—0 through 9. Since the combination has 3 digits, multiply $10 \cdot 10 \cdot 10 = 1{,}000$. There are 1,000 different possible combinations for the lock.

SKILLS TIP

The number of events in a given scenario is the number of numbers that are multiplied when using the Fundamental Counting Principle.

Using Probability Models

Probability

The **probability** of a certain event is the ratio of a given favorable outcome to the total possible number of outcomes. The formula for finding probability is shown here.

$$P(N) = \frac{N \text{ favorable outcomes}}{\text{total possible outcomes}}$$

For the combination lock in the previous example, what is the probability of randomly choosing the correct combination for the lock?

ANSWER: There is only 1 correct combination out of a total possible 1,000 combinations. So, the probability is

$$P(\text{correct combo}) = \frac{1}{1{,}000}$$

A four-digit personal identification number (PIN) is randomly generated to allow access to a website, using only the digits 0 through 9. What is the probability that the PIN is 1-2-3-4?

ANSWER: Each of the four digits in the PIN has 10 possibilities. So, the total number of possible outcomes is $10 \cdot 10 \cdot 10 \cdot 10 = 10{,}000$. Since only one combination results in the favorable outcome 1-2-3-4, the probability is

$$P(1\text{-}2\text{-}3\text{-}4) = \frac{1}{10{,}000}$$

Complete the activities below to check your understanding of the lesson content.

Questions 1 and 2 refer to the following information.

A quarter, a nickel, and a dime are each tossed, in that order. The possible outcomes of heads or tails are shown in the following tree diagram.

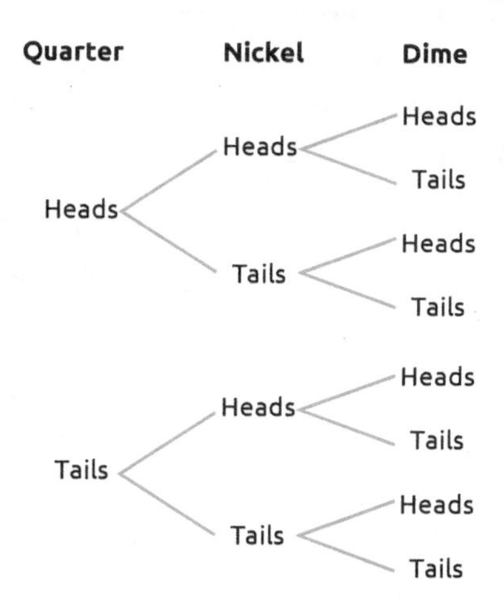

1. How many possible outcomes are there when flipping the three coins in order?

   A 2

   B 3

   C 4

   D 8

   E 14

2. What is the probability that all three coins land on tails?

   A $\frac{1}{2}$

   B $\frac{1}{4}$

   C $\frac{1}{8}$

   D $\frac{1}{12}$

   E $\frac{1}{16}$

3. Grantham has a bag full of letter cards—one card for each letter of the alphabet. He randomly selects two letters from the bag. What is the probability that he selects the letters O and then F if he does not replace the cards in the bag after each selection?

   A $\frac{1}{650}$

   B $\frac{1}{676}$

   C $\frac{1}{52}$

   D $\frac{1}{325}$

   E $\frac{1}{2}$

SKILLS TIP

If a tree diagram is used for probability, the last branch, farthest to the right, shows you the total number of possible outcomes in the scenario.

KEY POINT!

Probability is the ratio of the favorable outcomes to the total possible outcomes.

Lesson Practice

Questions 4 and 5 refer to the following information.

Each year, the NBA uses a lottery system to determine who will get the first pick in the draft. Ping pong balls numbered from 1 to 14 are placed in a machine, and four will be randomly selected. There are 1,000 possible combinations. All the teams not in the playoffs are given some number of combinations depending on their record—the team with the worst record gets the most combinations, and the team with the best record gets the least. The team with the combination that matches the one chosen by the machine will get the first pick in the draft.

TEST STRATEGY

Write the correct fraction, $\frac{250}{1,000}$, on scratch paper. If you see that it does not match any of the answer choices, check to see if it can be reduced or simplified.

4. If the team with the worst record in a season gets 250 combinations, what is the probability they will have the combination selected by the machine?

   A $\frac{1}{1,000}$

   B $\frac{1}{250}$

   C $\frac{1}{8}$

   D $\frac{1}{4}$

   E $\frac{1}{2}$

5. If the team with the best record in a season gets only 5 winning combinations, what is the probability they will win the first pick of the draft?

   A $\frac{1}{1,000}$

   B $\frac{1}{250}$

   C $\frac{1}{200}$

   D $\frac{1}{4}$

   E $\frac{1}{2}$

Questions 6 and 7 refer to the following information.

When selecting her outfit for the day, Helen chooses from three pairs of pants, five shirts, and two pairs of shoes that will coordinate well together.

6. How many possible different outfits can Helen create?

   A 3

   B 8

   C 10

   D 30

   E 60

7. What is the probability that Helen selects her blue pants, red shirt, and brown shoes?

   A $\frac{1}{60}$

   B $\frac{1}{30}$

   C $\frac{1}{10}$

   D $\frac{1}{3}$

   E $\frac{1}{2}$

See page 171 for answers and help.

Answer the questions based on the content covered in this unit.

1. Klem is giving hints to Cole about the 3-digit code to his lock. He tells Cole that the first digit is odd, the second digit is even, and the last digit is 2. If none of the numbers repeat, how many possibilities does Cole have to choose from?
   A 10
   B 20
   C 40
   D 50
   E 100

Questions 2 and 3 refer to the following table.

The class averages for Mrs. Simone's geometry classes are shown in the following table.

Period	Class Average (%)
1	88
2	90
4	84
6	86
7	72

2. What is the mean class average for all of Mrs. Simone's classes?
   A 16
   B 72
   C 84
   D 86
   E 90

3. Which period's class average might be considered an outlier? How might the measures of center be affected if the outlier is excluded? Explain your reasoning. Confirm your conclusion.

____

____

____

Questions 4–6 refer to the following table.

A local gym surveyed a sample of men and women ages 45–50 to better understand their exercise habits.

	Exercise 3 or more times per week	Exercise fewer than 3 times per week	Total
Men	21	14	35
Women	30	15	45
Total	51	29	80

4. How many people are included in the sample?
   A 80
   B 51
   C 45
   D 35
   E 29

5. According to the table, how many women exercise fewer than 3 times per week?
   A 80
   B 51
   C 45
   D 30
   E 15

6. According to the table, what percentage of men exercise 3 or more times per week?
   A 35%
   B 40%
   C 44%
   D 60%
   E 80%

Questions 7 and 8 refer to the following graph.

A music store owner took a random sample to determine the types of music customers like best.

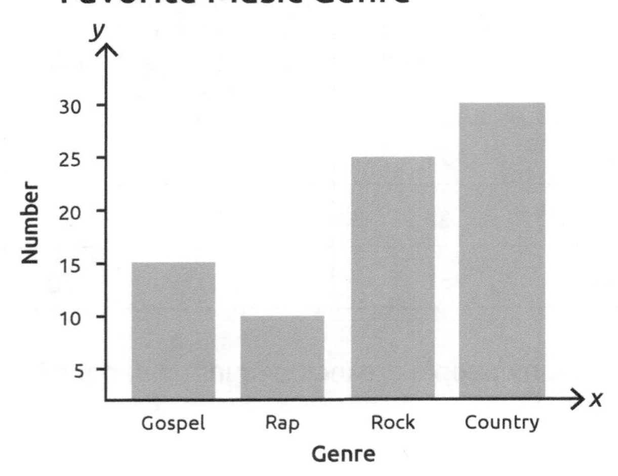

7. How many people are included in the sample?
   - **A** 15
   - **B** 30
   - **C** 80
   - **D** 100
   - **E** 200

8. According to the sample, about what percentage of the customers like gospel music best?
   - **A** 50%
   - **B** 33%
   - **C** 19%
   - **D** 15%
   - **E** 10%

9. Salvador rolls a six-sided die two times. What is the probability that he rolls two ones in a row?
   - **A** $\frac{1}{6}$
   - **B** $\frac{1}{12}$
   - **C** $\frac{1}{36}$
   - **D** $\frac{1}{48}$
   - **E** $\frac{1}{216}$

Questions 10 and 11 refer to the following box-and-whisker plot.

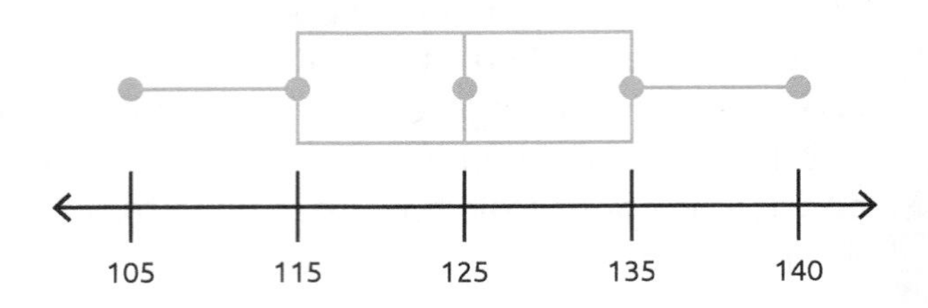

10. What does 135 represent on the box-and-whisker plot?
    - **A** 1st quartile
    - **B** median
    - **C** 3rd quartile
    - **D** maximum
    - **E** outlier

11. Which of the following is NOT true?
    - **A** About half of the data is above 125.
    - **B** All of the data is greater than 105.
    - **C** About half the data is between 115 and 135.
    - **D** About one-fourth of the data is greater than 135.
    - **E** About one-fourth of the data is less than 115.

Questions 12 and 13 refer to the following information.

The local senior center plans weekly recreational outings for its members. The planning team surveys a sample of members to find out their favorite activities. The center has 800 members.

Favorite Activity	Number of Members
Concert	16
Museum	11
Movie	25
Play Performance	28

12. How many people are included in the sample?
    - **A** 800
    - **B** 80
    - **C** 28
    - **D** 25
    - **E** 16

13. According to the sample, how many of the 800 members would likely prefer to attend a movie?
    - **A** 25
    - **B** 31
    - **C** 80
    - **D** 250
    - **E** 800

14. The scatter plot shows how time spent studying affects test grades. What type of relationship is shown in the scatter plot? Explain.

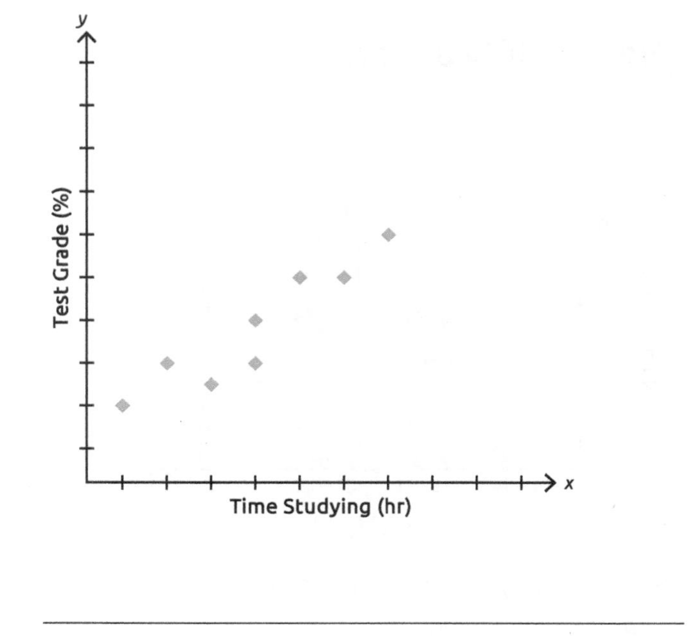

Questions 15 and 16 refer to the following information.

The following histogram shows the heights of students in a particular high school class.

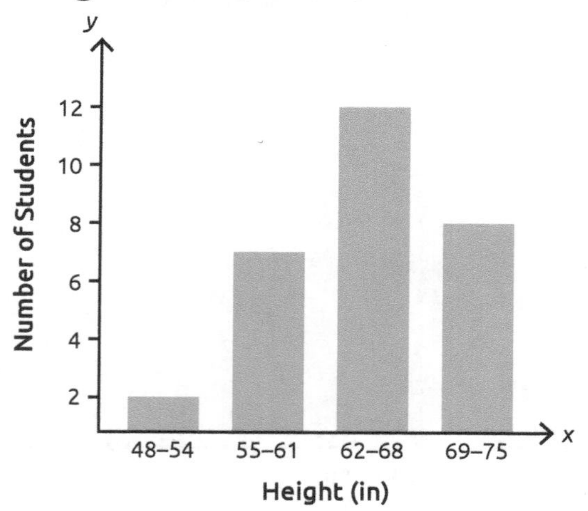

15. How many students are in the class?

A 2
B 8
C 12
D 29
E 75

16. How many students are at least 55 inches tall?

A 2
B 7
C 8
D 27
E 29

See page 171 for answers and help.

Questions 17 and 18 refer to the following information.

When selecting his lunch from the cafeteria, Mohammed chooses from four different sandwiches, three different types of fruit, and five different drinks.

17. If he wants a sandwich, a piece of fruit, and a drink, how many different three-item meal combinations can Mohammed choose?

A 5
B 12
C 20
D 60
E 120

18. What is the probability that Mohammed selects a turkey sandwich, an apple, and milk for lunch?

A $\frac{1}{60}$
B $\frac{1}{30}$
C $\frac{1}{10}$
D $\frac{1}{3}$
E $\frac{1}{2}$

Lesson 1

1. C. Divide the sum of all the capacities by 5.

2. B. Order the values from least to greatest and find the middle number.

3. A.

4. A decrease in the seating capacity at Stubhub Center will decrease the total seating capacity, which also decreases the mean seating capacity for all five stadiums.

5. E.

6. C. The median is better because the mean is influenced by the outlier.

Lesson 2

1. D. Add the number of people represented by all of the bars. $11 + 6 + 8 + 5 = 30$.

2. B. Divide the number of dog owners, 11, by the total number of people in the sample, 30.

3. C. About 27% of the sample own cats. Find 27% of 500.

4. It is unbiased because the customers are selected randomly and systematically.

5. C.

6. D.

7. E.

Lesson 3

1. E. The number in the bottom right cell is the sum of the people surveyed.

2. A.

3. C. The total in the Men's row shows the sum of the men who were interviewed.

4. The scatter plot shows a negative relationship. As the time practicing increases, the golf score decreases.

5. E.

6. B. The two intervals on the right represent the teams that had more than 39 wins.

7. A.

8. D.

9. C.

Lesson 4

1. D. The right-most branch of the tree diagram shows all of the possible outcomes.

2. C. All tails is 1 specific outcome out of the 8 possible outcomes.

3. A. The total possible outcomes are $26 \cdot 25 = 650$. There is one possible favorable outcome. So, the probability is $\frac{1}{650}$.

4. D. There are 250 favorable outcomes out of a total 1,000 outcomes. State the fraction in simplest terms.

5. C.

6. D. Multiply $3 \cdot 5 \cdot 2$.

7. B.

Unit Test

1. B.

2. C.

3. Period 7 might be considered an outlier. Excluding it would increase the mean and the median to 87%.

4. A.

5. E.

6. D.

7. C.

8. C.

9. C.

10. C.

11. B.

12. B.

13. D.

14. The scatter plot shows a positive relationship. As the time spent studying increases, the test scores increase.

15. D.

16. D.

17. D.

18. A.

Unit Glossary

- **bias** – occurs when one or more parts of a survey or sample are favored over others
- **box-and-whisker plot** – a graph that shows the spread of the data, using a number line and a box that display the median of the data and the first and third quartiles of the data
- **circle graph** – graph that displays data in sections of a circle
- **event** – a specific outcome or outcomes
- **frequency** – the number of times each data type occurs
- **Fundamental Counting Principle** – a way to determine the total number of outcomes possible by multiplying; it says that if event A has a possible outcomes and event B has b possible outcomes, then event A followed by event B has $a \cdot b$ total possible outcomes.
- **histogram** – a graph that displays numerical data that can be organized into equal intervals
- **inference** – a conclusion that can be drawn about a population based on evidence gathered from the random sample
- **interquartile range** – the range between the 1st and the 3rd quartiles of a data set
- **mean** – the sum of the numbers in the data set divided by the number of items in the set
- **measures of central tendency** – measures used to find the center of the spread of data—typically mean, median, and mode
- **median** – the middle number of the set after it is arranged in order from least to greatest; the average of the two middle numbers if the data set contains an even number of numbers
- **mode** – the number or numbers that appear most in the data set
- **outcome** – any of the possible results from a given action
- **outlier** – data points that are much larger or much smaller than most other values in a data set
- **population** – a larger group
- **probability** – the ratio of a given outcome to the total possible number of outcomes
- **quartile** – the numbers that divide the data into four equal parts
- **random sample** – a portion of the population that is selected in such a way that it represents the entire population
- **range** – the difference between the greatest number and the least number in the data set
- **sample** – a portion of a larger group
- **scatter plot** – a type of graph that can be used to demonstrate the relationship between two sets of data; the two sets of data are graphed as ordered pairs on a coordinate plane.
- **tree diagram** – a way to organize all of the possible outcomes

Understand Statistical Variability

- Know how to find weighted means.
- Know how to determine which measure of center is best for different sets of data.
- Understand the significance of outliers and how they can affect center and spread.

Use Random Sampling to Draw Inferences

- Know various types of sampling methods.
- Understand the difference between convenience and voluntary response sampling.

Describe Distributions

- Know how to read and interpret circle graphs.
- Know how to read and interpret line graphs.

Use Probability Models

- Know how to use probability models to make predictions and inferences.
- Know how to use permutations and combinations in probability.

Answer the following questions on the material you have learned.

1. Which of the following expressions represents the perimeter of the rectangle shown here?

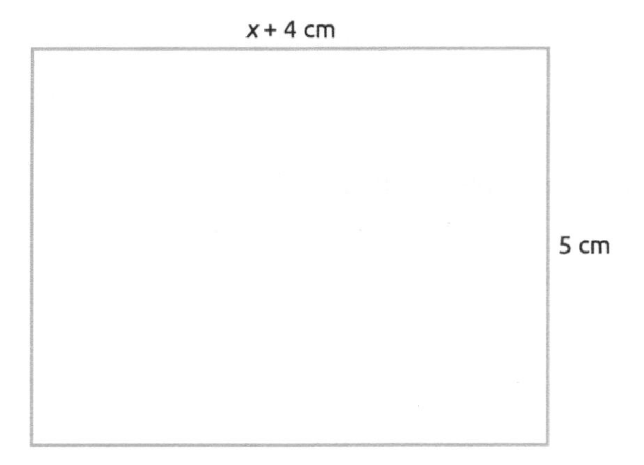

A $2x + 18$ cm
B $x + 9$ cm
C $5 + 2(x + 4)$ cm
D $8 + 2(x + 4)$ cm
E $5x + 20$ cm

2. A rectangular-shaped floor measures 96 feet by 68 feet. A flooring company is going to use ceramic tiles that each cover 2 square feet. Which of the following expressions represents the number of tiles that are needed for the floor?

A $\frac{2}{96 \times 68}$

B $\frac{96 \times 68}{2}$

C $\frac{2 \times 96}{68}$

D $\frac{2 \times 68}{96}$

E $68 \times 96 \times 2$

3. Which of the following graphs represents the line modeled by the equation $y = \frac{1}{4}x - 2$?

A
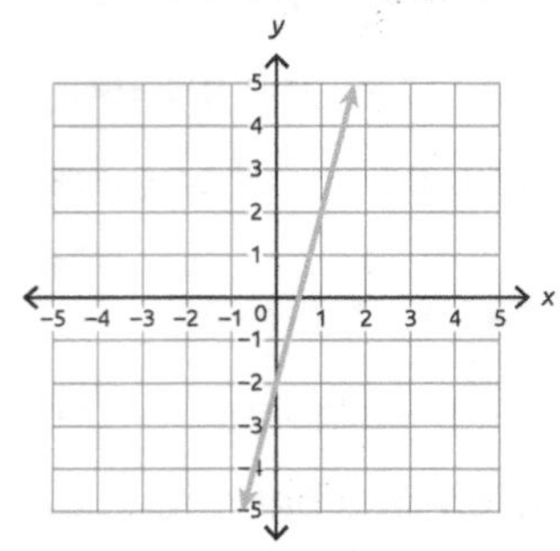

B
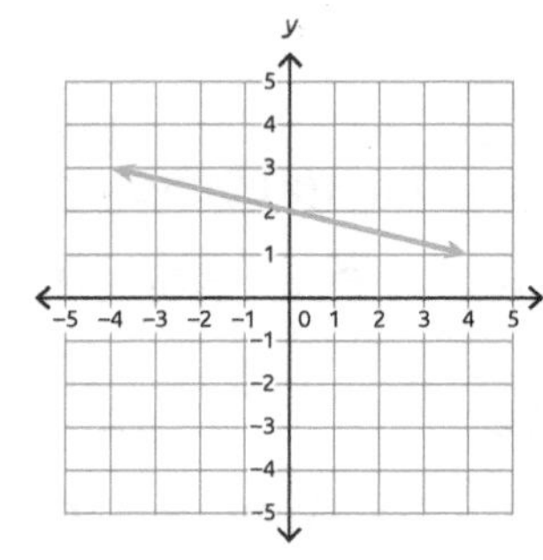

C
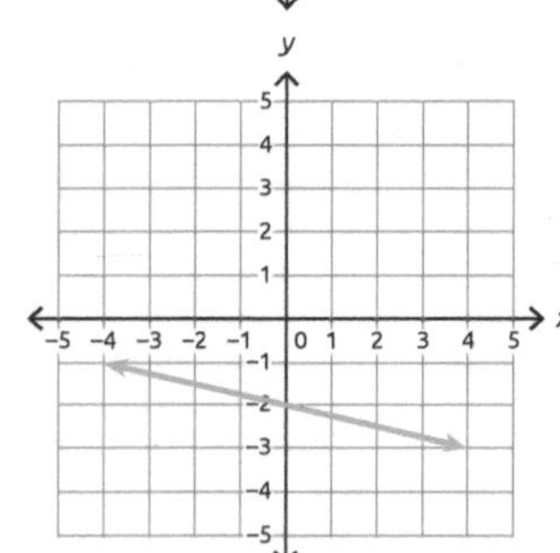

D
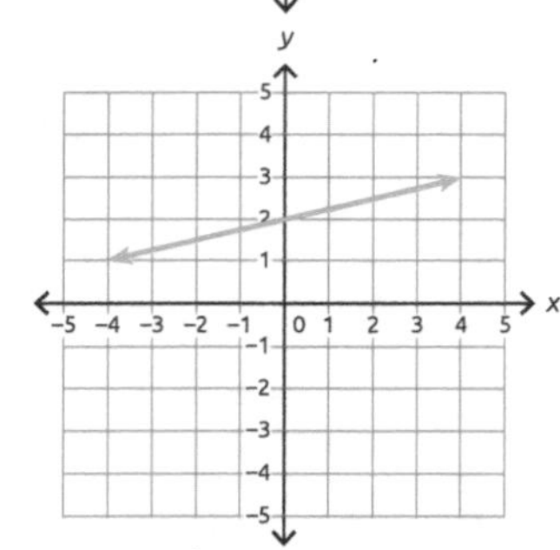

E
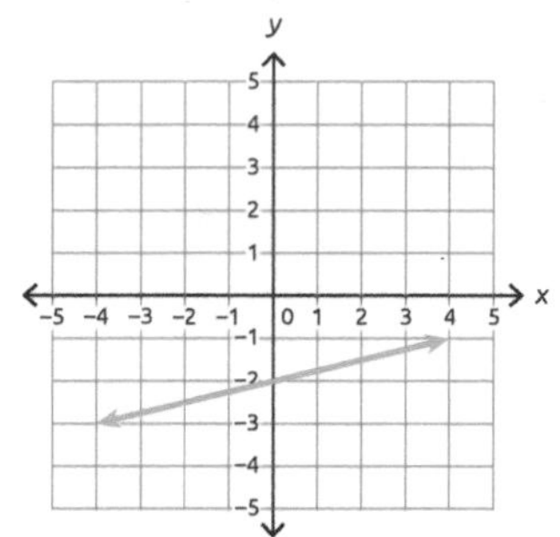

4. Which of the following statements is false?
   - **A** All integers are rational.
   - **B** All natural numbers are also whole numbers.
   - **C** A non-terminating, non-repeating decimal number is rational.
   - **D** All rational numbers are real.
   - **E** The number 0 is a whole number.

5. The store marks down the original cost of a shirt by 25% for a sale. Which of the following expressions represents what the store charges for the shirt that originally cost $22.50?
   - **A** $22.50 - 0.25(22.50)$
   - **B** $22.50 + 0.25(22.50)$
   - **C** $\frac{22.50}{0.25}$
   - **D** $0.25(22.50)$
   - **E** $\frac{22.50}{0.25(22.50)}$

6. A jug can hold 8 liters of water. If it fills at a rate of 50 milliliters per minute, how long will it take for the jug to fill up?
   - **A** 50 minutes
   - **B** 2 hours and 10 minutes
   - **C** 2 hours and 40 minutes
   - **D** 3 hours
   - **E** 3 hours and 40 minutes

7. Simplify the following expression completely.
   $\left(\sqrt{2} + 3\sqrt{3}\right)^2$
   - **A** 20
   - **B** 29
   - **C** $29 + 6\sqrt{6}$
   - **D** $31 + 3\sqrt{6}$
   - **E** $22 + 3\sqrt{6}$

8. Cameron has four shirts (red, blue, black, and white), two pairs of pants (navy and jeans), and three pairs of shoes (brown, black, and sneakers) to wear to work. What is the probability that Cameron will choose the red shirt, jeans, and sneakers to wear today?
   - **A** $\frac{1}{30}$
   - **B** $\frac{1}{24}$
   - **C** $\frac{1}{9}$
   - **D** $\frac{1}{3}$
   - **E** $\frac{1}{2}$

9. Which of the following expressions represents the perimeter of the rectangle shown?

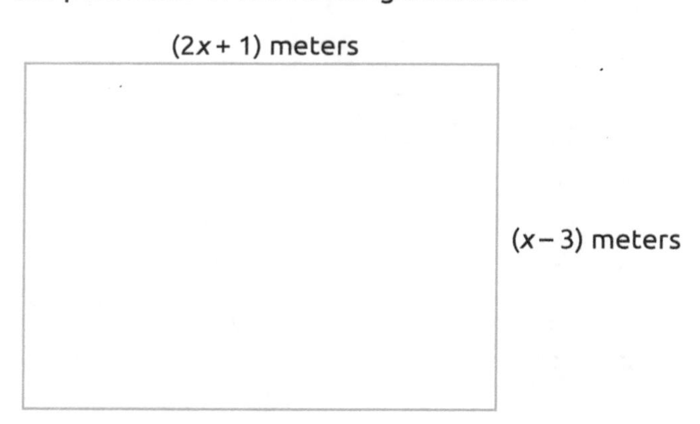

   - **A** $(3x - 2)$ meters
   - **B** $(4x + 1)$ meters
   - **C** $(6x - 4)$ meters
   - **D** $(2x^2 - 7x - 15)$ meters
   - **E** $(2x^2 - 13x - 15)$ meters

10. The bill for a hotel room was $136.50. Hal, Alvin, and Samaro agreed to divide the bill evenly among the three of them. Hal paid his portion with a $50 bill. How much change did he receive?

A $3.50
B $4.50
C $5.50
D $35.50
E $45.50

11. Jose is a quarterback for his football team. His first four rushes of the game are shown in the table. How many yards total has Jose rushed for so far?

Rush	Yards
1	5
2	−3
3	2
4	−1

A 11
B 3
C −1
D −3
E −11

12. On a basketball team, there are 7 guards, 5 forwards, and 2 centers. What is the ratio of guards to the number of players on the team?

A 1:7
B 5:7
C 1:2
D 2:7
E 7:5

13. Andrew is shopping for the best price on flour for his pastry shop. He compares prices from 5 different local and online stores. The table shows the advertised prices he found. Which store has the lowest price per pound of flour?

Store	lbs	Price
A	3	$10.20
B	2	$7.50
C	5	$11.25
D	10	$28.50
E	4	$17.00

A Store A
B Store B
C Store C
D Store D
E Store E

14. Theresa's parents loaned her $5,000 at 4% annual simple interest for 3 years. At the end of the loan, how much money will she repay?

A $600
B $4,200
C $4,400
D $5,600
E $5,800

15. Nat has three packages of ground beef that weigh 6.65 lbs, 4.85 lbs, and 12.5 lbs. He wants to divide the beef into six equal packages. How much ground beef will be in each package?

A 2 lbs
B 3 lbs
C 4 lbs
D 6 lbs
E 12 lbs

16. Which of the following expressions gives x yards, y feet, and z inches in terms of feet?

 A $x + y + z$

 B $3(x + y + z)$

 C $x + 3y + 36z$

 D $3x + y + \frac{1}{12}z$

 E $\frac{1}{3}x + y + 12z$

Questions 17 and 18 refer to the following information.

Tony earns $250 for each car he sells. He also earns a base salary of $500 per month.

17. If Tony earned $4,500 last month, which of the following equations represents the number of cars he sold?

 A $250x - 4{,}500 = 500$

 B $250x + 4{,}500 = 500$

 C $250x + 500 = 4{,}500$

 D $250x - 500 = 4{,}500$

 E $500x - 250 = 4{,}500$

18. If Tony earned $4,500 last month, how many cars did he sell?

 A 8

 B 11

 C 13

 D 16

 E 17

19. Which of the following uses the Distributive Property to help find the product of 9 and 27.6?

 A $9 + 27 \times 9 + 0.6$

 B $9(20) + 9(7) + 9(0.6)$

 C $9(27) \div 9(0.6)$

 D $9(27) - 9(0.6)$

 E $9(27) + 9(7.6)$

Questions 20 and 21 refer to the following information.

Luis is driving from his home in Dallas to San Antonio. He stops at a rest area after driving 165 miles, which is three-fifths of the way.

20. Which of the following equations can be used to determine the distance from Dallas to San Antonio?

 A $\frac{3}{5} \div x = 165$

 B $\frac{3}{5}x = 165$

 C $\frac{3}{5} \cdot 165 = x$

 D $\frac{3}{5} + x = 165$

 E $x - \frac{3}{5} = 165$

21. How many total miles does Luis have to drive to reach San Antonio?

 A 99 miles

 B 110 miles

 C 165 miles

 D 275 miles

 E 300 miles

22. Solve the following equation.

 $5 + \sqrt[3]{5x + 7} = 8$

 A 81

 B 64

 C 27

 D 4

 E no solution

Questions 23 and 24 refer to the following information.

There are three consecutive integers. The sum of the three integers is equal to 84.

23. If n is the smallest integer, which of the following represents the integers?

 A $n+n+1+n+2=84$

 B $n+n+2+n+4=84$

 C $(n+n+1)-2(n+1)=84$

 D $2(n+n+1+n+2)=84$

 E $2(n+n+1)=84$

24. What are the three consecutive integers?

 A 26, 27, and 28

 B 26, 28, and 30

 C 27, 28, and 29

 D 27, 29, and 31

 E 28, 30, and 32

Questions 25 and 26 refer to the following figure.

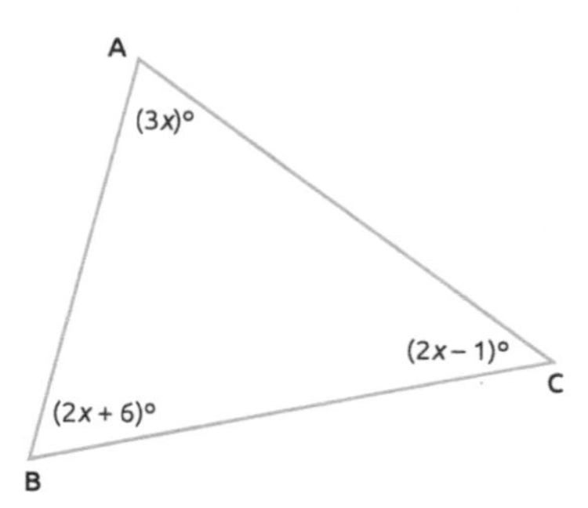

25. In ΔABC, what is the value of x?

 A 20

 B 25

 C 52

 D 83

 E 180

26. In ΔABC, what is $m\angle B$?

 A 25°

 B 49°

 C 56°

 D 75°

 E 180°

27. The area of a rectangle can be found using the formula $A = lw$. Which of the following represents the area of the rectangle shown?

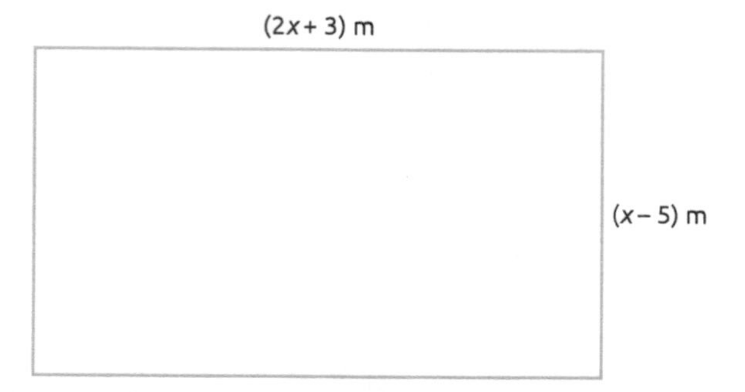

 A $(3x - 2)$ m²

 B $(4x + 1)$ m²

 C $(6x - 4)$ m²

 D $(2x^2 - 7x - 15)$ m²

 E $(2x^2 - 13x - 15)$ m²

Questions 28 and 29 refer to the following information.

Jamison has 19 copies of 2 baseball cards that are valuable to collectors. Card A is worth $30, and Card B is worth $45. The total value of all the cards is $735.

28. Which of the following is a system of equations that represents the number of card A and card B?

 A $\begin{cases} 30A+45B=19 \\ A+B=735 \end{cases}$

 B $\begin{cases} 30A+45B=735 \\ A+B=19 \end{cases}$

 C $\begin{cases} 30A+45B=735 \\ 30A+45B=19 \end{cases}$

 D $\begin{cases} A+B=735 \\ A+B=19 \end{cases}$

 E $\begin{cases} 19A+19B=735 \\ A+B=19 \end{cases}$

29. How many \$30 and \$45 cards does Jamison have?

   A 7 \$30 cards and 12 \$45 cards
   B 11 \$30 cards and 8 \$45 cards
   C 10 \$30 cards and 9 \$45 cards
   D 9 \$30 cards and 10 \$45 cards
   E 8 \$30 cards and 11 \$45 cards

30. Which inequality is represented in the following graph?

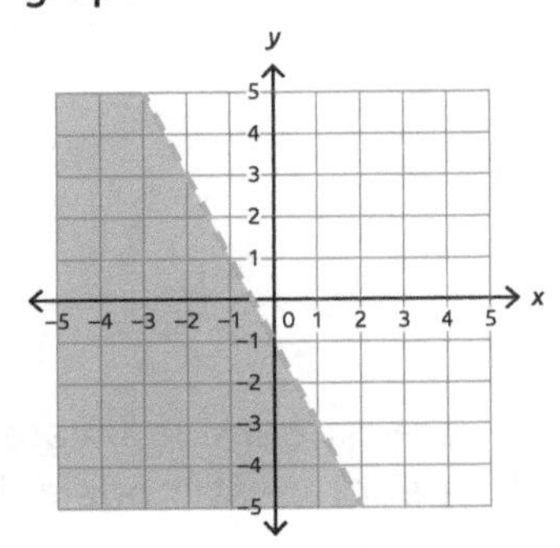

   A $y \leq -2x - 1$
   B $y < 2x - 1$
   C $y < -2x - 1$
   D $y > -2x - 1$
   E $y \leq -2x + 1$

31. The area of a square can be found using the formula $A = s^2$, where s is the length of a side of the square. What is the area of the square shown here?

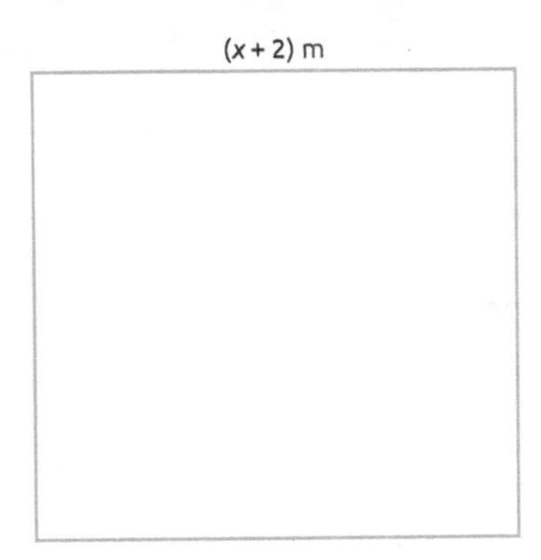

   A $(x^2 + 4)$ m²
   B $(x^2 + 4x + 2)$ m²
   C $(x^2 + 4x + 4)$ m²
   D $(2x + 4)$ m²
   E $(4x + 8)$ m²

Questions 32 and 33 refer to the box-and-whisker plot below.

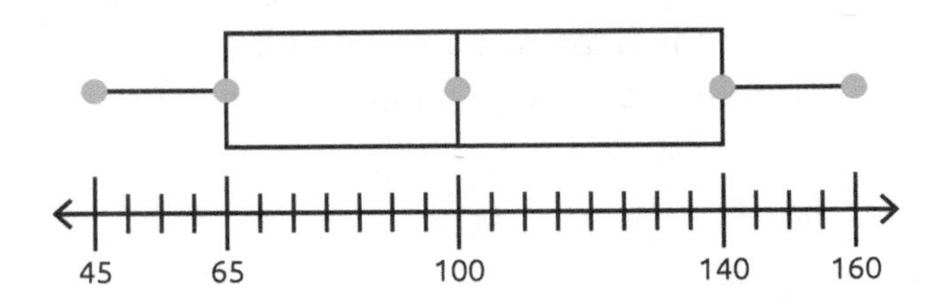

32. What does 160 represent on the box-and-whisker plot?

   A 1st quartile
   B median
   C 3rd quartile
   D maximum
   E outlier

33. In this box-and-whisker plot, the region inside the box represents data between which two values?

   A the minimum and the maximum
   B the first quartile and the second quartile
   C the first quartile and the third quartile
   D the second quartile and the third quartile
   E the median and the maximum

34. Consider the equation $4x + 5 = 8x - 5$. Which of the following can be the first step in solving this equation?

   A. Add 5 to both sides of the equation.
   B. Subtract 4x from both sides of the equation.
   C. Divide both sides of the equation by 4.

   A A only
   B B only
   C C only
   D A and B only
   E B and C only

35. To convert temperature from Celsius to Fahrenheit, use the formula $F = \frac{9}{5}C + 32$, where F is degrees Fahrenheit and C is degrees Celsius. To the nearest tenth, what is the temperature in degrees Fahrenheit when the temperature is 32°C?

   A −32°F
   B 0°F
   C 57.6°F
   D 89.6°F
   E 112.6°F

36. What is the solution of the following system of equations?

$$\begin{cases} -4x - 6y = -24 \\ 2x + 3y = 12 \end{cases}$$

   A (3, 2)
   B (0, 4)
   C (6, 0)
   D There is no solution.
   E There are infinitely many solutions.

Questions 37 and 38 refer to the following graph.

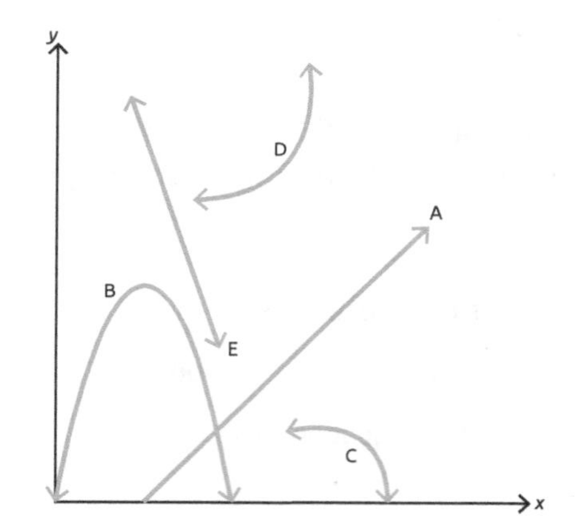

37. According to the shape and the direction of the graph, which figure in the graph could correspond to the function $f(x) = -(1.05)^x + 120$?

   A A
   B B
   C C
   D D
   E E

38. Which figure in the graph above models a quadratic function?

   A A
   B B
   C C
   D D
   E E

39. If $f(x) = -5x - 2$, what is $f(-5)$?

   A −27
   B −23
   C 8
   D 23
   E 27

40. The picture shows the water stream of a water fountain. What type of function best models the flow of water from the fountain?

   A linear
   B quadratic
   C exponential
   D linear and exponential
   E linear and quadratic

41. If $f(x) = 5x - 1$, which of the following is $f^{-1}(x)$?

A $f^{-1}(x) = -5x + 1$

B $f^{-1}(x) = 5x + 1$

C $f^{-1}(x) = \frac{1}{5}x + \frac{1}{5}$

D $f^{-1}(x) = \frac{1}{5}x + 1$

E $f^{-1}(x) = \frac{1}{5}x - 1$

Questions 42 and 43 refer to the following information.

The following graph shows two transformations of triangle ABC.

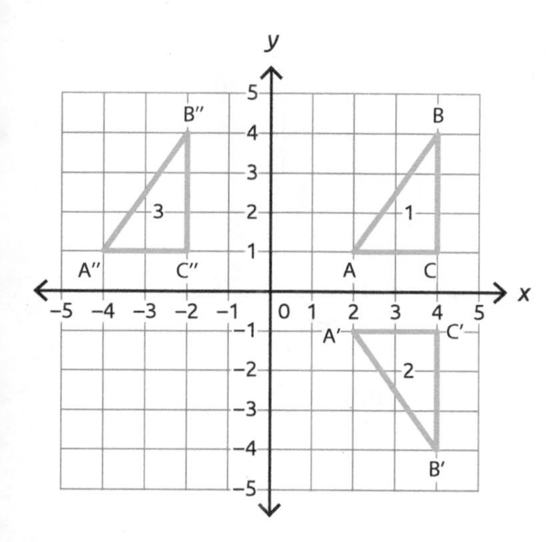

42. Which transformation takes ΔABC onto $\Delta A'B'C'$?

A translation

B rotation

C reflection

D dilation

E dissection

43. Which transformation takes ΔABC onto $\Delta A''B''C''$?

A translation

B rotation

C reflection

D dilation

E dissection

44. Leonard is striping the lines of a rectangular athletic field on a grass surface. He wants to make sure that the corner at angle Z is square. What value must x be to ensure this?

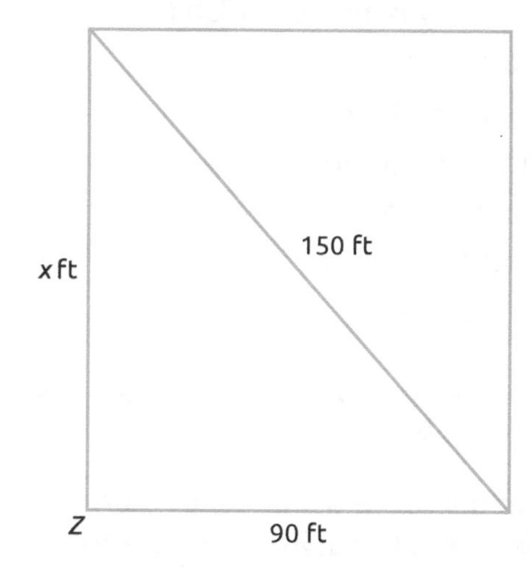

A 60

B 90

C 120

D 125

E 145

45. Using the values in the table, which of the following expressions represents the approximate number of inches in 4 miles?

12 inches = 1 foot
3 feet = 1 yard
5,280 feet = 1 mile

A 2.5×10^9

B 2.5×10^7

C 2.5×10^5

D 2.5×10^{-5}

E 2.5×10^{-7}

46. Students in field and track are practicing the long jump. The coach is recording the lengths of the jumps to the nearest $\frac{1}{100}$ of a meter. If a student's jump is recorded as 4.60 meters, the student's jump was actually between which two amounts?

    A 4.55 and 4.64 meters

    B 4.595 and 4.604 meters

    C 4.50 and 4.65 meters

    D 4.50 and 4.604 meters

    E 4.595 and 4.6 meters

Questions 47 and 48 refer to the results of the survey, shown in the table below.

A car dealership surveyed a sample of men and women to understand the style of vehicle they like best.

	Luxury Car	Sports Car	Total
Men	62	98	160
Women	135	25	160
Total	197	123	320

47. How many people are included in the sample?

    A 123

    B 160

    C 197

    D 320

    E 960

48. Write this number in standard notation: 9.87×10^{-5}

    A 0.0000987

    B 0.000987

    C 0.00987

    D 0.0987

    E 0.987

49. According to the table, what percentage of the men liked luxury cars best?

    A 19.375%

    B 30.625%

    C 31.5%

    D 38.75%

    E 61.25%

50. The average times for the members of the track team running the 400-meter race are tabulated and displayed in the histogram shown. How many students can run 400 meters in less than 58 seconds?

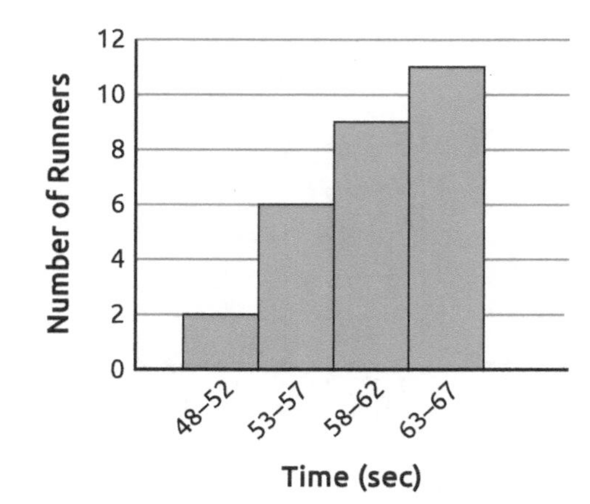

    A 2

    B 6

    C 8

    D 17

    E 28

See page 183 for answers and help.

Practice Test Answer Key

1. **A.** $2(x + 4) + 2(5) = 2x + 8 + 10 = 2x + 18$

2. **B.** Multiply the area of the floor 96×68 and divide by 2.

3. **E.**

4. **C.**

5. **A.**

6. **C.** There are 8,000 mL in the jug. 8,000 divided by 50 is 160. Convert 160 minutes to hours and minutes.

7. **C.** Multiply the binomial by itself: $\left(\sqrt{2} + 3\sqrt{3}\right)\left(\sqrt{2} + 3\sqrt{3}\right)$.

8. **B.** He has $3 \times 4 \times 2 = 24$ possible outfits to choose from. So, the probability is 1 out of 24, or $\frac{1}{24}$.

9. **C.**

10. **B.** Divide \$136.50 by 3 and subtract the quotient from \$50.

11. **B.** Add $5 + (-3) + 2 + (-1) = 3$.

12. **C.**

13. **C.** Find the price per pound at each store by dividing the price by the number of pounds.

14. **D.** Multiply \$5,000(0.04)(3) to find the interest, and then add that product to the original loan amount, \$5,000.

15. **C.**

16. **D.**

17. **C.** Multiply the number of cars he sold, x, by 250. Add the salary, \$500, to that, which makes the sum \$4,500.

18. **D.**

19. **B.**

20. **B.**

21. **D.** Solve:

$$\frac{3}{5}x = 165$$
$$\frac{5}{3} \times \frac{3}{5}x = 165 \times \frac{5}{3}$$
$$x = 275$$

22. **D.**

23. **A.**

24. **C.** Combine like terms and solve the equation: $n + n + 1 + n + 2 = 84$

$$3n + 3 = 84$$
$$3n = 81$$
$$n = 27$$

25. **B.** The sum of the three angles is 180. Set up the equation $3x + (2x - 1) + (2x + 6) = 180$, and then solve for x.

26. **C.**

27. **D.** Multiply the polynomials: $(2x + 3)(x - 5) = 2x^2 - 10x + 3x - 15$ $= 2x^2 - 7x - 15$.

28. **B.** Set up a value equation where \$30 times A plus \$45 times B equals \$735. Then, since there are 19 cards altogether, A plus B equals 19.

29. **E.**

30. **C.**

31. **C.** Multiply $(x + 2)(x + 2)$.

32. **D.**

33. **C.**

34. **D.**

35. **D.** $F = \frac{9}{5}C + 32$

$$= \frac{9}{5}(32) + 32$$
$$= 57.6 + 32$$
$$= 89.6° \text{ F}$$

36. **E.**

37. **C.**

38. **B.** The graph of a quadratic function is a parabola.

39. **D.** $f(-5) = -5(-5) - 2 = 25 - 2 = 23$.

40. **B.**

41. **C.** Replace $f(x)$ with y. Swap x and y, and then solve for y:

$$y = 5x - 1$$
$$x = 5y - 1$$
$$x + 1 = 5y$$
$$\frac{1}{5}x + \frac{1}{5} = y$$

42. **C.** The triangle is flipped over the x-axis, which is a reflection.

43. **A.** The transformation is a slide 6 units to the left.

44. **C.** Use the Pythagorean Theorem to find the value of x.

$$x^2 + 90^2 = 150^2$$
$$x^2 = 150^2 - 90^2$$
$$x^2 = 14{,}400$$
$$x = 120 \text{ ft}$$

45. **C.**

46. **B.**

47. **D.** The bottom right cell in the table is the sum of all the people in the sample.

48. **A.**

49. **D.** Divide the number of men who stated they like luxury cars best, 62, by the total number of men in the sample, 160.

50. **C.**

Use the following formulas as reference when working on problems in this book.

Perimeter/Circumference
Rectangle *Perimeter = 2(length) + 2(width)*
Circle *Circumference* $= 2\pi$*(radius)*

Area
Circle *Area* $= \pi(\text{radius})^2$
Triangle *Area* $= \frac{1}{2}$ *(base)(height)*
Parallelogram *Area = (base)(height)*
Trapezoid *Area* $= \frac{1}{2}(\text{base}_1 + \text{base}_2)(\text{height})$

Volume
Prism/Cylinder *Volume = (area of the base)(height)*
Pyramid/Cone *Volume* $= \frac{1}{3}$ *(area of the base)(height)*
Sphere *Volume* $= \frac{4}{3}\pi(\text{radius})^3$

Length
1 foot = 12 inches
1 yard = 3 feet
1 mile = 5,280 feet
1 meter = 1,000 millimeters
1 meter = 100 centimeters
1 kilometer = 1,000 meters
1 mile ≈ 1.6 kilometers
1 inch = 2.54 centimeters
1 foot ≈ 0.3 meter

Capacity/Volume
1 cup = 8 fluid ounces
1 pint = 2 cups
1 quart = 2 pints
1 gallon = 4 quarts
1 gallon = 231 cubic inches
1 liter = 1,000 milliliters
1 liter ≈ 0.264 gallon

Weight
1 pound = 16 ounces
1 ton = 2,000 pounds
1 gram = 1,000 milligrams
1 kilogram = 1,000 grams
1 kilogram ≈ 2.2 pounds
1 ounce ≈ 28.3 grams